Boo
iot.
ecall
4.2.

TURN

7 JA   RN 998
                                        4

DUE   99   IN
2 FE   N

RETU   RN
AR   ED

F   of borrowi

13 MAR 1995

Optical fibre sensing and signal processing

Optical fibre sensing and signal processing

B.Culshaw

Peter Peregrinus Ltd.
On behalf of the Institution of
Electrical Engineers

Published by: Peter Peregrinus Ltd., London, UK.

© 1984: Peter Peregrinus Ltd.

British Library Cataloguing in Publication Data

Culshaw, B.
   Optical fibre sensing and signal processing.
   1. Optical communications   2. Fibre optics
   I. Title
   621.38'0414   TK 5103.59

   ISBN 0-906048-99-0

Printed in England by Short Run Press Ltd., Exeter

Contents

Preface

The potential of optical fibres in applications other than communications has been recognised only relatively recently. Consequently this is a subject area which is expanding rapidly and which includes an ever-increasing range of technological nuances. In writing this book, I have attempted to ignore the majority of the nuances and concentrate on the most important optical principles which underlie the operation of these particular systems. In particular, the intricacies of mechanical, chemical and materials aspects of sensing systems have been largely omitted, and the details of signal processing functions and formats have also been covered far more competently by other authors.

The primary objectives of this book are to bring together into one volume the relevant optics and to illustrate the application of these optics to a range of optical fibre sensors and optical signal processors. A range of sensing concepts is discussed, from the simplest intensity modulated devices through to phase- and polarisation-sensitive transducers in which the optics involved becomes relatively complex. Signal processing using fibre optics is an even younger science, and the chapter devoted to this topic describes some basic concepts and makes an attempt to put this technology into context.

Optical fibre sensing and signal processing is a multidisciplinary activity, embracing many — if not all — aspects of conventional optical fibre communications together with a great deal of fundamental optics. The details of all these topics have been competently covered by many specialist authors, and the treatment in this volume makes frequent reference to these related texts. This book seeks to fulfil a need which I felt when starting to work with optical fibre systems for non-communications applications — that is, to provide the introduction to the relevant physics in other texts and to present a basis for the development of a reliable physical feel for the underlying principles. The approach is therefore to develop straightforward, relatively simple arguments rather than to enter into considerable academic detail, which often obscures the development of physical intuition. In addition, a full and detailed treatment would run to several volumes, which I would never have completed and you the readers would never have purchased, and which would have involved the repetition of much standard material!

Countless people have contributed to the development of this book. Students and colleagues past and present at University College London, on whose labours much of this book has been built, must start the list. Then there are the discussions with countless individuals from industry, university and government, who between them have contributed to forming a picture of the role that these systems may play. The actual writing of this book was done while I was on sabbatical leave from University College London at Stanford University. I am extremely grateful to both UCL for the opportunity to spend some time at Stanford, and to Stanford for helping it all to happen. In particular, John Shaw and Chap Cutler at Stanford, together with other staff and students at the Ginzton Laboratory, have contributed greatly, especially in the signal processing and fibre gyroscope aspects of the book. My thanks go also to Mavis Small for translating the text into legible type with patience and understanding. Finally, the greatest of thanks are due to my wife Pat and daughter Shauna for tolerating the author.

Brian Culshaw
Mountain View
July 1982

Chapter 1

Expanding the role of fibre optics

1.1 Introduction

Communication using light is an ancient art. The North American Indian developed an efficient and adequate communications network based on what we would now term pulse code modulation of a subcarrier, and semaphore and flashlight signalling have been in use for centuries. Measurement using light is almost as old. The optical lever has been around for over a century — a simple example being the mirror galvanometer — and remote chemical analysis using light (Fraunhofer lines) was discovered a considerable time ago.

In all these systems, the hand often did the optical modulation, the eye did the optical detection, and the light was transmitted through air. Air can, of course, have a remarkably low loss. But it fluctuates continually in its transmission characteristics; thus the stars twinkle and often disappear completely. The human eye has many excellent features as a very complex high-resolution optical detector, but it is slow, which is very fortunate for the television industry. The human hand leaves much to be desired as an optical modulator. But optical systems remained in this basic form for a considerable time, while a complex and extensive electronic industry developed in the background. The change came in 1966 when Kao and Hockham, working from STL in Harlow, England, published a paper demonstrating that light may be guided with low loss in thin, flexible glass fibre guides [1.1]. This opened up new vistas. It became feasible to consider guiding light over many kilometres, thereby eliminating the uncertainties of atmospheric transmission. Fifteen years later, the industry had become established, and lightwave communications has become one of the important cornerstones of the telecommunications industry.

This has obviously done little directly for our optical lever, but already there is the possibility of having a light source — at the end of a fibre — which is readily repositioned without realigning the source optics. There is also the possibility of collecting light from the optical lever and transmitting this somewhere else for analysis. So simple optical sensors with fibre feeds become a relatively straightforward adaptation of the basic optical measurement technique, but with the advantages of a remote source and detection system.

Meanwhile, in the mid 1970s it began to be appreciated that the fibre itself may form the basis of a new direct transduction principle, interfacing the field to be measured with the light guided within a fibre without any intervening stages [1.2, 1.3]. The concept of the all-fibre sensor has since then become reality in numerous laboratories throughout the world.

Optics has other interesting properties. In particular, it can be used as the basis of numerous information processing systems, and these rather elegant ideas have been turned into practice on optical benches in most — if not all — developed countries [1.4, 1.5]. Optical systems may also be assembled to perform many memory and switching functions analogous to the familiar gate structures in electronics. There is a wide variety of basic optical building blocks waiting to be exploited.

But why the interest in optics? Not least among the reasons is the inherent fascination of the technology and the background physics. This stimulates the researcher, but there have to be practical possibilities to encourage his sponsor. In communications these include long-distance transmission of large data rates, elimination of ground loop problems, compact reliable transmission paths and low cost per unit transmission bandwidth. In sensing there are other attractive features, including immunity to electromagnetic interference, low thermal and mechanical inertia for the sensing material and the use of chemically inert materials for the sensing head. In addition, the sensor consumes zero electrical power, and, particularly in future installations, there is a considerable potential for a flexible system format, possibly incorporating signal processing into the basic sensor.

Optical fibres in signal processing offer the means to guide high-bandwidth signals over long distances with low signal dispersion. This suggests that delay lines may prove to be an important element in fibre optics. Fibres may also play a part in modifying some of the optical bench processors into a rugged format aimed at manipulating spatial input and output functions.

This is a brief background to the material in this book. Optics in general, and fibre optics in particular, is responsible for a vast and rapidly expanding literature [1.6]. This indicates, correctly, that this is a subject area which is changing extremely rapidly. This observation has influenced the content of this short volume, which is limited to gathering together a number of basic principles pertinent to the noncommunications applications of fibre optics and related technologies, and to illustrating these principles by referring to a number of sensing and signal processing systems. The objective is to establish the technological and conceptual framework rather than expound on the details of the numerous individual, and often idiosyncratic, realisations of these concepts. The hope is that the basic principles will be explained and that sufficient application of these principles is provided to enable the reader to evaluate his or her own particular problem. The book may be divided into three sections; these are the principles, the practice and the possibilities of fibre optic technology in sensing and signal processing.

1.2 The principles

Perhaps a better title for the book would have been *Guided Lightwaves in Sensing and Signal Processing,* since the scope of the book includes integrated optics, though fibre optics dominates. Integrated optics – the optical frequency analogue of a microstrip circuit in microwaves – has considerable potential in both applications, and any discourse omitting this important technology would be incomplete. There are many texts on integrated optics, so that the subject will be treated here from an applications aspect rather than a technological aspect [1.7]. But one part of explaining the principles is to describe the basic features of dielectric waveguides used in both fibre and integrated optics.

The remainder of the section of principles (Chapters 3–5) is devoted to a discussion of sources and detectors used in sensing and signal processing systems, and finally to an analysis of the means available for modulating and demodulating an optical carrier wave. The approach is that of a system designer rather than a device manufacturer, and the objective of the discussion is to establish the factors influencing the fundamental limits on sensitivity and resolution in all the systems discussed. Most of this section is the application of fairly standard optical fibre communications concepts to the sensing and signal processing application. There are some aspects of the subject which are peculiar to this application, in particular with respect to optical sources. The subject also relies considerably on many of the concepts of classical optics – far more so than conventional optical communications systems – and so appendixes are included to very quickly describe the essential optics. These appendixes will serve as a reminder to those exposed to optics in the past and as a guide to the relevant areas in the classical optics texts for those new to the subject.

1.3 The practice

There is an infinity of ways in which guided light may be modulated by an environmental parameter. However, one can define a general optical fibre sensor, and Fig. 1.1. shows a schematic diagram of such a device. A constant light source (constant can mean constant in intensity, frequency, phase, colour, polarisation, or several of these) is launched into an optical fibre to a region in which the light is modulated in one of the aforementioned constant properties. The light is then returned from the modulation zone along another fibre to be detected and demodulated.

There are two basic classes in fibre sensor. The first is one in which the fibres serve as a light source and light collector and the modulation process takes place externally from the fibre, usually by way of an attenuation process which is modulated by the measurand. The attenuation may be implemented by moving masks or reflectors or by indirect processes. These indirect processes can include variable-birefringence crystals causing polarisation modulation, which is converted to intensity modulation by an analyser; the exploitation of chemical absorption spectra; the use of scintillations caused by the interaction of the modulator zone with radioactive particles; and the use of phosphorescence and luminescence phenomena. The scope for intensity modulated transducers is clearly very wide ranging, and is often as much a problem in materials science as in optical fibre technology.

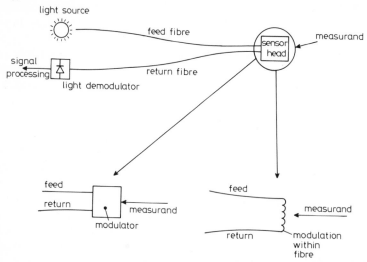

Fig. 1.1   *General features of an optical fibre sensor showing the distinction between extrinsic (externally modulated) and intrinsic sensors*

The second class of sensor is that in which the measurand interacts directly with the light in the fibre. Phase, polarisation and intensity may all be modulated within the fibre, using appropriately designed fibre and cable. The first of these is probably the most common. The advantages of this class of sensor stem from the fact that there are no optical interfaces at the modulator head. The disadvantages stem from the fact that, in general, if the fibre within the modulator head is capable of imposing modulation of the light passing within it, so too are the feed and return fibres. This difficulty – which has been dubbed the 'lead sensitivity' problem – has not yet been satisfactorily overcome. In some of these sensors, the problem does not arise. For instance, in the optical fibre gyroscope these effects are self-cancelling by virtue of the design of the gyroscope interferometer, and in some versions of the fibre hydrophone the device may be useful as a single high-sensitivity detector which is self-contained and has electrical feeds and returns. These sensors have the highest sensitivity of optical fibre sensors, and a number of solutions to the lead sensitivity problem are emerging. This will open up the potential for highly sensitive optical fibre remote sensors with the capability of multiplexing several of these on to one fibre link.

Chapters 6–11 are devoted to detailed discussions of the basics of a variety of sensors. The examples chosen to illustrate the discussion are intended to demonstate the principles involved. In all cases there are very many other sensors which may be configured based on the same principles.

Signal processing using fibre optics is in its infancy at the time of writing the book (mid-1982). However, it is possible to identify three groups of optical signal processors in which fibre optic components could play a central part. Spatial optical processors have been studied for approaching a quarter of a century (see, for instance, Reference 1.4). The way in which fibres could assist in the exploitation of this technology is probably limited, but a few of the possibilities are discussed in the signal processing chapter (Chapter 12). Integrated optics has considerable potential in spatial processing. Nonlinear optics is interesting in that direct analogues of electronic gate structures are feasible, and extremely fast switching processes are available. Fibre optics is an attractive concept here, since the required power densities may be more readily reached in a guided mode. However, it is probably fair to comment that radical advances in materials technology are required before the necessary optical power levels (typically in the order of several gigawatts per square centimetre) are reduced to be compatible with the construction of compact and complex high-speed programmable switching arrays. The technological competitor here is probably the Josephson junction device; this has fundamental advantages and, though it is cryogenic, many of the concepts have been successfully demonstrated [1.8]. The final, and most appropriate, application for guided wave optics is in delay line signal processing. Optical fibres have potentially the highest time-bandwidth product of any delay medium known by orders of magnitude. Tapping this reservoir of signal processing capability presents a number of technological problems, and some of the early attempts are described in Chapter 12.

1.4 The possibilities

Chapter 13 maps out some of this endless variety. Optical sensing and signal processing, when coupled with fibre optic transmission and the refinements of the associated technologies, is without doubt a powerful tool for an enormous range of applications. During the course of the book we will have established the basic principles and identified a number of current, viable applications for the techniques. But with optical technology at its present stage, these applications are restricted – to high electromagnetic interference zones, remote locations, hazardous atmospheres, certain medical applications and some research laboratory detection systems. The available optical and materials technology is, however, expanding at a significant rate, so that the components from which sensing and signal processing systems may be built will by 1990 include, among others, elegant monomode fibre systems and integrated optics. The options open for the systems designer will then increase dramatically. The final chapter in the book indulges in some speculation as to the types of sensor and signal processing systems which will be feasible in the relatively near future. The reason for including this chapter is to emphasise that the technology is young and that the possibilities are boundless. Any limits over and above basic physics — which are covered in this book — are imposed only by the imagination and initiative of the designers involved. But this last chapter should serve to indicate the tremendous potential which is locked into optical techniques. There is the capability, both economic and technical, for completely modifying the design philosophy for industrial and military signal processing and sensing philosophies in the foreseeable future.

Chapter 2

Principles of optical fibres

2.1 Introduction

This chapter presents a short introduction to the principles and properties of optical fibre waveguides. The treatment is brief and aimed at achieving a basic understanding of the important optical phenomena in fibres, with particular reference to the application of fibres in sensor systems. Much of the material in this chapter may also be applied to integrated optic waveguides, which operate on very similar principles but with more complex fabrication processes and different symmetry properties in the guiding structure.

There are many texts to which the interested reader may be referred for more detailed treatment of the material in this chapter. For example Midwinter's book [2.1] presents a most useful account of the underlying physics of optical fibres. In addition there have been a number of reprint volumes; the most recent and comprehensive is edited by Kao [2.2], and gives an indication of the current status of optical fibre systems and the present research trends in this rapidly developing field.

The emphasis in most treatments of optical fibres is on applications in the PTT (post, telegraphs and telephones) industries. Transducer technology and signal processing applications require different transmission properties, and often the impressive performance of the PTT fibres is unnecessary though a useful bonus. The demand from PTT is likely to be substantial, and it is probably realistic to operate under the premise that the use of PTT fibres in transducer and signal processing applications will be the most cost-effective approach.

This chapter describes first the principles of fibre waveguides with cylindrical symmetry, discusses their transmission properties, especially with respect to dispersion and attenuation, and then examines the essential features of fibre components — that is, connectors, splices, splitters etc. Finally, largely by way of completing the background on fibres, some discussion of the present trends in optical fibre systems is included.

2.2 Principles of optical fibre waveguides

The basic ray optical principles of an optical fibre waveguide are shown in Fig. 2.1.

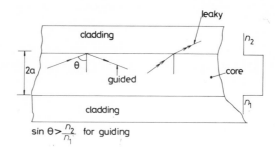

Fig. 2.1   *Longitudinal section of a step index optical fibre showing guided and leaky rays*

A dielectric rod of refractive index n_1 is enclosed in a dielectric tube of lower refractive index n_2. A ray of light propagating along the rod will be totally internally refracted at the interface if

$$\sin\theta > n_2/n_1 \qquad\qquad (2.1)$$

simply by applying Snell's law at the interface [2.3]. If this inequality is not met, then the light will leak from the central region of the fibre to the surrounding glass, and will be lost. In the case of an optical fibre guide, the central region is referred to as the fibre *core* and the surrounding tube as the *cladding*. Rays meeting eqn. 2.1 are *guided rays*, and the remainder are *leaky rays*.

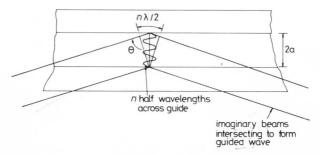

Fig. 2.2   *Showing the relationship between allowed ray directions and the electric field across the guide*

The core radius a in most optical fibre guides is in the region of 2 to 200 microns, and so the fact that light is a wave motion with a wavelength in the order of 0.5 to 1 microns is relevant in determining the properties of this waveguide. Thus, any light path within the guide must satisfy boundary conditions imposed by applying Maxwell's equations at the interface between the core and the cladding. The consequence of this is that there are specific 'allowed' ray directions which correspond to setting up electric fields across the guide to satisfy the boundary condition. The essential requirement for this is shown in Fig. 2.2. In the simplest case, that' of a rectangular guide with boundary conditions of zero electric field at the core/ cladding interface, the allowed ray directions satisfy:

$$\cos \theta = \frac{m\lambda}{2 \cdot 2a} \tag{2.2}$$

where m is the number of half wavelengths of electric field across the guide. The number m is one of the *mode numbers* of the propagating mode; the second mode number is determined by satisfying similar boundary conditions in the orthogonal direction across the guide.

In the case of this hypothetical guide, satisfying eqn. 2.1 for the propagation angle and eqn. 2.2 for the mode numbers, we see that there is a maximum value of the mode number, which occurs when $\sin \theta = n_2/n_1$. With a little manipulation, it is easy to show that:

$$m_{max} = \frac{4a}{\lambda} \left\{ 1 - \left(\frac{n_2}{n_1} \right)^2 \right\}^{1/2} \tag{2.3}$$

This relationship is most important in optical guides, since it serves as a criterion to determine whether a given guide is single mode or multimode. If m_{max} is less than 2 then the guide will be *single moded;* otherwise the guide will be *multimoded.* Note that for single-mode operation the guide core width will be of the order of a few optical wavelengths. Equation 2.3 can be expressed in terms of more usual guide parameters – the refractive indices are usually not specified. However, the *numerical aperture* of the guide is always known, and this is defined by the maximum launch angle at which light is guided. Fig. 2.3 shows the basic geometry. The numerical

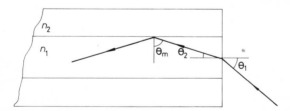

Fig. 2.3   *Derivation of numerical aperture*

aperture NA is $\sin \theta_1$ in the diagram. Hence:

$$NA = \sin \theta_1 = n_1 \sin \theta_2$$

$$= n_1 \left\{ 1 - \left(\frac{n_2}{n_1} \right)^2 \right\}^{1/2} \tag{2.4}$$

$$= (2n\Delta n)^{1/2}$$

where n is the mean refractive index of core and cladding and Δn is the refractive index difference. Eqn. 2.3 can now be expressed as:

$$m_{max} \simeq \frac{4a}{\lambda_0} NA \qquad (2.5)$$

where λ_0 is the free-space wavelength.

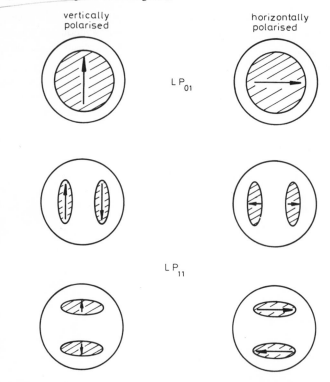

vertically
polarised

horizontally
polarised

LP_{01}

LP_{11}

Fig. 2.4   *Diagram of field distributions for LP_{01} and LP_{11} modes*

An optical fibre waveguide is circular in cross-section, and the boundary condition at the core/cladding interface involves a finite field. However, the overall conclusions of the preceding analysis are modified only in detail. The condition for single-mode operation is:

$$a \leqslant \frac{2\cdot4\lambda}{NA} \qquad (2.6)$$

and the field distributions across the guide are now cylindrically symmetric and may be specified by two mode numbers, a radial number m and an azimuthal number l. It is often convenient to think of a given mode as one of a linearly polarised set (LP modes), and some example field distributions for low-order LP modes are shown in Fig. 2.4. Apart from the LP_{01} mode there are four field distributions associated with a given mode number, corresponding to $\sin l\phi$ and $\cos l\phi$ in the azimuthal direction and to horizontal and vertical polarisations. The general form of the LP mode distribution is:

$$E(r, \phi) = F_m(r), \cos l\phi \tag{2.7}$$

where E is the electric field and F is the function required to satisfy the boundary conditions at the core/cladding interface and the dielectric constant distribution within the core region. Detailed accounts of this are given in Reference 2.1.

If the guide is perfectly circularly symmetrical, then the fields for the various forms of the LP modes are exactly identical and so these modes will propagate with exactly the same velocity (i.e. the modes are degenerate). However, fibres are never exactly circular, and so there is a slight difference in the field distributions and the modes are no longer exactly degenerate. In most cases this is of no consequence. There is, however, an important exception, which is the case of the nominally single-moded guide, and this is particularly relevant to a large class of optical fibre sensors – the interferometric ones. Light injected into a single-mode fibre in different polarisations will propagate at different phase velocities; consequently the output polarisation is usually not related to the input, and also varies with factors such as temperature and the mechanical condition of the fibre.

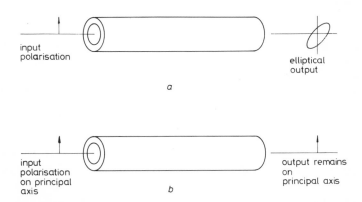

input polarisation

elliptical output

a

input polarisation on principal axis

output remains on principal axis

b

Fig. 2.5   *Retention of polarisation states using birefringent fibres*
(a) circular fibre
(b) elliptical fibre

The subject of polarisation in single-mode optical fibres may be distilled into the statement that the birefringence of a section of fibre may be reduced into two components – linear and circular (see Appendix 1). Thus linearly polarised input light will, in general, emerge from the linear birefringent section elliptically polarised, and then the plane of the ellipse will be rotated in the circular birefringent section. If the linear input light is along the principal axis of the linear component, then the emerging light will be linearly polarised but rotated. The effective linear birefringence is, for most fibres, a function of the temperature and mechanical conditions of the fibre, and the circular birefringence a function of the degree of mechanical twist of the fibre. The variations in the linear birefringence stem largely from the fact that, in single-mode fibre, significant mode coupling occurs between the two orthogonally polarised modes, induced by mechanical perturbations. This is basically because the propagation constants of the two modes are very close to each other. However, there is a class of optical fibres, in general with elliptical symmetry in the core [2.4, 2.5] induced either by index variations or through the physical size of the core in which the propagation constants of the two principal modes are deliberately made very different. In these fibres, light launched in a linear polarisation along one of the principal axes emerges along this principal axis, still linearly polarised, but possibly rotated from the input plane (see Fig. 2.5).

2.3 Properties of optical fibres

The properties of optical fibres may be divided into two groups, the electrical aspects — that is, the attenuation and dispersion of signals passing along the fibres — and the mechanical and geometrical properties, which determine environmental effects on the fibre and the ease with which light may be launched into the core.

2.3.1 Attenuation

The total attenuation of light passing along an optical fibre depends on three factors, Rayleigh scattering, absorption due to impurities (especially water) and microbending due to loss induced by the mechanical condition of the fibre.

Rayleigh scattering imposes the lower limit on achievable attenuation in optical fibres. The scattering is due to minute (small compared to the optical wavelength) inhomogeneities in the core material. These are, in turn, a direct consequence of the cooling process during the drawing of the glass fibre from the preform rod. The size of the inhomogeneities is directly related to the temperatures attained during the drawing process. Rayleigh scattering varies as $1/\lambda^4$, and so increases dramatically as the wavelength decreases. In high-quality optical fibres, the observed attenuation is very close to the Rayleigh limit at wavelengths of approximately one micron and below.

Absorption lines due to the presence of impurities in the silica fibre are most noticeable for water, since there is a water (strictly, OH) absorption peak at a wavelength of 1·38 microns. By keeping impurity levels low, typically in the order of parts in 10^9, this absorption peak can be kept to the same order as the Rayleigh scattering. There is a much more detailed account, including the effects of other important impurities on these phenomena, in Chapter 8 in Midwinter (2.1).

In a well constructed optical fibre, Rayleigh scattering will dominate the loss process. There is a subtle, but important, distinction between the Rayleigh scattering process and impurity absorption processes. In absorption processes the incident light energy is captured by the impurity atoms and becomes thermal energy — consequently heating the core of the fibre. Rayleigh scattering (which is, of course, the phenomenon responsible for the blue sky) is not in itself an absorption of energy; it is a spatial redistribution of the light incident upon these minute discontinuities in the core glass structure. The differences are illustrated in Fig. 2.6. Some of the scattered light is guided in the core, both as forward and backscatter. It is relatively simple to estimate the magnitude of this guided backscattered radiation. Scattering produces reradiation of a small percentage of the incident energy over a full 4π solid angle. Of this, that in the solid angle defined by the numerical aperture of the fibre will be guided backwards. Thus a fraction $(NA)^2/4$ of the scattered energy returns to the source, and the same amount is also forward scattered. The amount of backscattered radiation is thus readily related to the fibre loss. The backscattered radiation plays an important role in some optical sensors, especially the gyroscope.

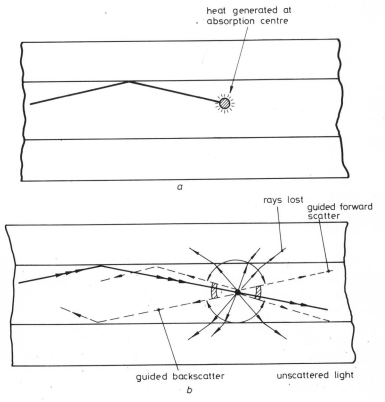

Fig. 2.6　*Power loss due to (a) absorption and (b) scattering in optical fibres*

It may also be used as a measuring tool (in time domain reflectometers) to indicate both the levels of Rayleigh scattering and the location of discontinuities in the fibre propagation path by examining the time variation of the returned backscattered radiation [2.6].

Present-day, high-quality optical fibres exhibit attenuation characteristics very close to the limits imposed at short wavelengths by Rayleigh scatter and at longer wavelengths by the presence of an intrinsic absorption line in the core material itself in the far infra-red. Most fibres show a residual OH absorption peak at 1·38 microns, and the resultant attenuation spectrum will be typically as shown in Fig. 2.7 (there have been reports of complete elimination of the OH peak in very carefully controlled fibres [2.7]).

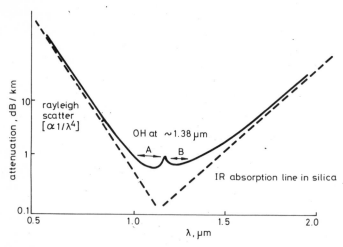

Fig. 2.7   *Typical attenuation dependence on wavelength for a high-quality optical fibre*

There are two transmission windows in this type of fibre, from roughly 800—1300 nm and from 1450 nm to 1600 nm (arbitrarily defined as the region in which attenuation is less than 3 dB/km). The current communications band is the 850—950 nm region, set by the availability of reliable and economic sources and detectors. There is currently much effort on systems in the 1300 nm and 1500 nm range, since here the attenuation is noticeably lower, and the dispersion may also be reduced (see below). Single-mode systems are also more readily achieved at these wavelengths since mechanical tolerances are considerably relaxed.

In principle, attenuation figures even lower than the fractions of 1 dB/km obtained in silica fibres may be realised by using different core materials. This is a topic currently in its infancy, but attenuation less than 10^{-3} dB/km is predicted with longer wavelengths still (4 to 8 microns) to reduce Rayleigh scatter and using fibres consisting of halide glasses [2.8].

These attenuation figures may be readily destroyed by inducing microbending losses during the cabling process or during fibre handling. The art of fibre optic cable making has now progressed to the stage where excess losses induced in cabling are usually negligible. However, it is important to appreciate the principles of microbending-induced losses, in part because this forms the basis of a number of sensor devices.

Microbending loss is simply expressed as mechanically induced coupling between modes which are guided in the core to modes in the cladding, which are thereby lost from the core. This loss is enhanced when the spatial period of the mechanically induced perturbation along the fibre coincides with the difference in wave numbers of adjacent modes within the fibre. This difference is typically of the order of milli- metres, so that spatially induced coupling may readily occur. For instance, tightly winding a drum of 3 dB/km fibre may increase the losses of the same fibre to over 20 dB/km. Thus cabling techniques are designed to ensure that minimal mechanical strain is transmitted from the cable structure to the fibre it contains.

2.3.2 Dispersion

If a sharp pulse of light is injected into an optical fibre, it emerges from the other end somewhat wider, owing to dispersion along the fibre. The dispersion of a fibre is usually specified in terms of the pulse broadening per kilometre of fibre path. There are two sources of dispersion within the fibre itself; these are due to inter- mode dispersion and material dispersion.

Intermode dispersion is a direct consequence of the fact that in a multimode fibre (or even in a single-mode fibre, in which there are two orthogonal polarisation modes) the modes travel with different velocities. Thus an impulse of light which launches power equally into all modes will emerge at the end of the fibre distributed in time, owing to the physical path differences of the various modes.

For multimode fibre, this difference may be minimised by the use of graded index profiles (usually appproximately parabolic) whereby the time delay for various modal paths may be, in principle, equalised (see Fig. 2.8). Typically, a step index fibre will exhibit a dispersion of the order of tens of nanoseconds per kilo- metre (ns/km) and a graded index fibre of the order of 1 ns/km.

Material dispersion is that occurring on a single mode owing to the variation in refractive index of the core glass with wavelength. The total dispersion then depends on both the wavelength and the bandwidth of the source. The material dispersion is expressed in picoseconds of pulse dispersion per kilometre of fibre path per nanometre of source linewidth (ps/(km nm)). Typically, this varies with wavelength as shown in Fig. 2.8. Note that there is a zero at about 1·3 microns, so that a single-mode fibre using this propagation wavelength would exhibit zero first- order pulse dispersion (there will, of course, be second-order effects to make the total dispersion small but finite). The position in wavelength of the zero dispersion may be tuned by designing a waveguide whose dispersion characteristics are opposite in sign to the small dispersion levels around the 1·3 micron wavelength zero in the material dispersion. There is currently much activity in this area for long-distance communications applications.

The dispersion for a given source/fibre combination is obtained by taking the square root of the sum of the squares of the material and intermode dispersions. In determining the former, due note should be taken of the total frequency spread of the source during the modulation cycle. For instance, a semiconductor laser may have a small bandwidth at a constant current level, but the bandspread introduced when its current and/or temperature is modulated will generally greatly exceed the stable linewidth unless the laser is mode locked (Reference 2.9). It should also be noted that, in particular, intermode dispersion in multimode fibres should be evaluated with care, since the measured value depends on the mode spectrum launched and on the amount of mode mixing which occurs along the fibre path. Detailed accounts of these phenomena may be found in Midwinter [2.1].

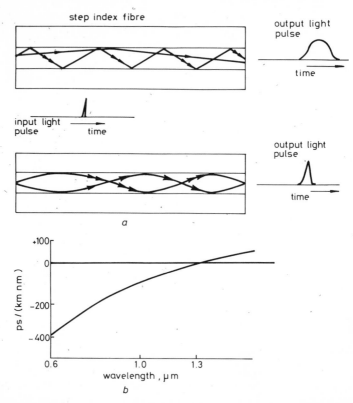

Fig. 2.8   *(a) A ray model demonstrating intermode dispersion in multimode fibres. (b) A typical variation of material dispersion with wavelength for silica-based fibres. The exact curve depends on the doping in the material*

2.3.3 Mechanical properties

The mechanical strength of optical fibres is intrinsically very high. However, early fibres were found to exhibit very poor mechanical characteristics, due to, it was later discovered, the existence of microcracks on the fibre surface which grew in the presence of moisture until they travelled through the entire fibre. This particular effect is now completely controlled by the addition of a primary coating. This coating, usually a silicone resin, is added during the drawing process and completely seals the surface against moisture ingress. The complete fibre drawing process is then as shown — very diagrammatically — in Fig. 2.9.

Optical fibres are usually relatively thin strands, typically less than 200 microns, and as such may be bent into very small radius of curvature without significant chance of breakage. For many fibres, a bending radius of less than 1 cm is easily

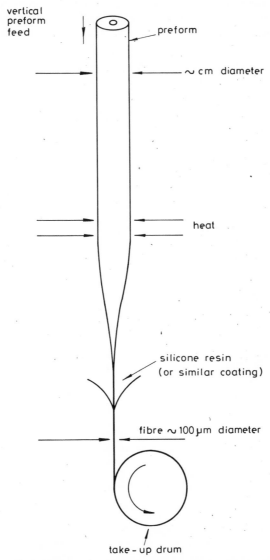

Fig. 2.9   *The essential features of fibre drawing from preforms. Alternative processes — for instance, double-crucible or continuous deposition techniques — are also available*

achieved without any damage to the fibre. However, for thicker fibres (fibres with cladding diameters of more than 500 microns are available) bending imposes a proportionately greater stress on the surface on the outside of the curve, and breakages are likely for bending radii of less than about 10 cm, especially if the fibre is likely to be subjected to pronounced thermal fluctuations.

There is one other point about bending, namely that bending induces a loss by encouraging conversion of the propagated light within the core to cladding modes. In general, bending radii of less than 10 cm will induce significant bending losses, though there are great variations depending on the detailed structure of the fibre. Bending also plays an important part in certain single-mode systems. The differential stresses introduced by the bending process produce a linear birefringence with its principal axes parallel to and perpendicular to the radius r of the bend. This effect can be used as a polarising component in single-mode systems, though the fact that the extent of the linear birefringence is proportional to $1/r^2$ means that the effect is only important as a spurious source of birefringence in compact systems or in systems where there is a considerable length of wound single-mode fibres – for instance in delay line signal processors and in the fibre optic gyroscope.

Optical coupling between fibres and from optical sources to fibres depends totally on achieving a match between the size and numerical aperture of the core and that of the source. These effects are shown schematically in Fig. 2.10. It is possible to achieve some limited matching between source NA and dimensions and cores NA and dimensions using lens structures. However, it should be noted (see Apendix 1) that for simple optical systems the product (object area $\times (NA)^2$) will remain a constant.

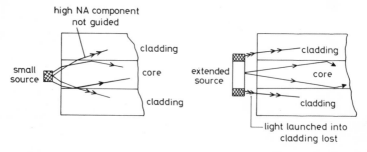

Fig. 2.10   *Illustrating the importance of numerical aperture and area matching between source and fibre*

2.4 Optical fibre components

Component technologies for a wide variety of applications of optical fibres are available. Here we include connectors and splices, and couplers, combiners and splitters.

In connectors and splices the same basic mechanical alignment criteria discussed in the previous paragraphs apply, but usually with the additional constraint that the fibres to be coupled must be identical in both NA and diameter for efficient connection. In some connectors a lens structure is used, in which case the more general constrains apply. It is a straightforward but lengthy process to calculate the losses caused by offsets in the alignment of the two cores, variations in core diameter and variations in *NA*. These are considered in Chapter 13 of Reference 2.1.

It is appropriate here to make a few general, basic points about fibre connectors and splices. First, the smaller the core, the more difficult the tolerancing required for successful connectors and splices. There is, therefore, a widely available inexpensive connector technology for large-core (over 100 microns) fibres, but connectors are expensive and require skilled assembly for smaller-core fibres (50 microns and less). The technology for single-mode fibre connectors is correspondingly more difficult, though numerous laboratory models have been proved for use in the 1.3 and 1.5 micron wavelength band.

Splicing has only received serious attention for use with PTT grade fibres (cores of 50 microns and below) and is now developed into a repeatable field-operable science. The particular attraction of splices over connectors are that the former have a lower loss and are mechanically more tolerant of environmental variations. Splices involve V-groove techniques or rod techniques to align the fibres to be joined, and the final jointing is completed by either an index matched epoxy or by using spark fusion or a microtorch. Some form of alignment procedure is usually followed to ensure that the coupling between the two fibres is optimum, but surface tension effects will align the claddings if the splice involves melting the fibre. This inevitably requires access to one other end of one of the fibres being jointed (either for backscatter or transmission measurements). Splicing is rarely considered for larger-cored fibres. These are invariably high-dispersion step index fibres with much higher attenuations levels than in PTT fibres. Connectors are used for input and output from the fibre, which is usually short (less than 500 metres) in total transmission length.

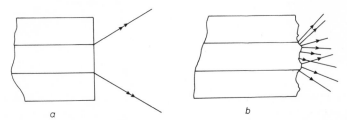

a   *b*

Fig. 2.11   *The effect of poor end face preparation (a) plane fibre end, output light circularly symmetrical (b) scratched and/or cracked fibre end, output light not symmetrical*

Both splicing and connector mounting processes require very careful preparation of the end faces of the optical fibre. The end should be both perpendicular to the axis of the fibre and have a surface which is optically flat. Fortunately these requirements are often met by a simple cleaving process, though for the larger-core fibres a polishing step may be included in the connector mounting procedure. The flat end face is required in connectors to ensure that the minimum amount of light is scattered on leaving the end face into directions which cannot be guided in the next section of fibre (see Fig. 2.11). In splicing, the region between the end faces to be jointed is usually either flooded with index matching liquid, or the glasses are fused. However, end face preparation is still important to ensure that the cores are accurately aligned before the joint is made permanent.

Couplers, splitters and combiners are readily available for large-core-diameter fibres − 200 micron region − and in practice it is usually in applications requiring larger-diameter fibres (for example, short-haul interconnected data systems) that the splitter and combiner are more frequently required. Up to about a 16-way splitter is at present available, and research devices with up to 30-way splits have been reported. For smaller-core multimode fibres, the system requirement is lower and the problem more difficult. PTT systems are invariably point to point transmission routes. In monomode fibres, the coupler is now a 'wave' device (rather than a 'ray' device as for 200 micron core fibres). The constraints on a single-mode coupler are then that the propagation constants in the two sections be equal and that the propagation length and coupling coefficient be correct for the desired coupling. A single-mode coupler is then, in all senses, a microwave device, but operating at micron wavelengths. Some devices have been built, and in many cases the availability of a usable single-mode coupler is central to the operation of a sensor or signal processing system. The use of couplers will be discussed along with the relevant systems [2.10].

2.5 Optical fibres − discussion

A definite pattern has evolved in the types of optical fibre which are currently available (spring 1982). These may be classified as low-performance step index fibres, PTT graded index multimode fibres, and monomode fibres.

The PTT fibre is in the greatest demand, and is the only one which has approached anywhere near a standardised form. The general consensus among a wide variety of world-wide manufacturers is towards a 50 micron core, 125 micron cladding diameter parabolic index guide with a dispersion in the region of 1 ns/km. There is a substantial and continuing demand for this type of fibre, which is probably manufactured in greater length than all other types of fibre together. Even so there remains a connection problem, in that each manufacturer has his own unique solution to the design of a demountable connector, invariably incompatible with the design offered by his competitors. Joints and splices also vary considerably between manufacturers, though all fall into either the fusion splice category or the expoxyed splice category. The differences are in details imposed by the fact that there are differing cabling philosophies and designs, falling into the so-called 'loose tube' and 'solid core' forms. Both designs are used to overcome the substantial penalties which can accrue due to microbending loss in improperly cabled fibres. Taps, couplers and splitters are almost unknown in PTT systems, so there is very little available. However, there is some considerable work in progress on wavelength multiplexing in PTT systems (largely aimed at local distribution networks), in which sources of several wavelengths are combined together via a multiplexer to travel along one path and then split at the receive end (see Fig. 2.12). These systems will typically involve the transmission of about four channels along the same fibre.

Step index fibres are offered in a bewildering variety of types and sizes. The applications of step index fibres are usually in short-haul, low-data-rate communications links, and so attenuation and dispersion are no longer prime performance criteria. Glass and silica fibres are available, the latter with both silica cladding and in the form of PCS (plastic-clad silica) fibre. PCS fibre consists of a solid silica core with a silicone resin cladding which both forms the guide and protects the outer surface. There is to date no consensus standardisation of step index fibres, though considerable efforts are being made by both the BSI in the UK and the IEC internationally to produce recommended standards. Many inexpensive connector components are available for step index fibres, though there is no standard form of connector. There is also a limited range of splitting and combining networks.

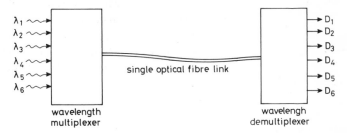

Fig. 2.12   *The principles of a wavelength multiplexed system*

Monomode fibres are now at the 'prototype systems' stage, in that there are a number of operators (PTT again) who have installed trial communications links using transmission sources in the 1.3 and 1.5 micron wavelength bands. These systems are still very new, and much has to be learnt about their behaviour. At shorter wavelengths (850 nm), there is a more mature knowledge of the basic systems parameters, but less experience in handling the proportionally smaller single-mode structures.

Where do applications to sensors and signal processing fit into this scene? Sensors, as will become much more apparent later in the book, may be divided into three very broad groups. The first comprises those using the fibre as a light pipe to convey illumination to a remote point. Here the simplest of step index fibres is usually adequate. The second group consists of those using the fibre to convey light to a remote point at which the light is modulated in phase or polarisation. In this case the properties of the fibre between the source and the sensor are critical, and special fibres, invariably single mode, are often required (Reference 2.11 describes one such special fibre). In the final group the fibre itself may be used as the sensing element as well as the transmission element. Again it is often, but not exclusively, the case that special fibre constructions will be required.

Signal processing using fibres is still a young science, but again one could make some broad generalisations concerning the applicability of various fibre types. Fibre delay lines are compact and have a wide time-bandwidth product. They can be effectively realised in graded index multimode fibres (PTT quality). Even higher performance, in terms of time—bandwidth product, is achievable using mono-mode. Still other signal processing concepts involve the use of monomode guides with switching, combining or logic components. These applications involve the use of compact, long delays. Thus the fibre should be designed to be used at very tight bend radii. The bending losses are minimised by the use of relatively high numerical apertures, and the fibres are often metal coated: the normal silicone-based primary coating is vulnerable when subjected to prolonged differential stresses of the magnitudes expected on fibres wound in radii of the order of 1 cm.

This chapter has but scratched the surface of the topic of optical fibres, with the objective of providing a very elementary understanding of the topic as back-ground for the appreciation of the use of fibres in sensor and signal processing systems. The subject develops rapidly, and there are, each year, numerous con-ferences describing recent advances. The interested reader is referred to these proceedings as well as the texts mentioned earlier for a full and more complete treatment of the topic (see Reference 2.12).

Chapter 3

Optical sources for sensors and signal processing

3.1 Introduction

Optical fibre systems have used almost every conceivable light source from flash-light bulbs to high-power lasers. In this chapter, we briefly examine the properties and principles of those which are most commonly used. The chapter is a very general treatment of the subject, but it should perhaps be mentioned that although some technologies are changing rapidly, many general issues − for instance, spatial matching criteria between source and fibre − remain.

The science of optical sources is constantly changing. However, in the context of optical fibre systems it is possible to define a set of objectives for the performance characteristics of the source. The question is then whether a suitable source exists which is compatible with these objectives. A basic knowledge of source properties is then necessary to ascertain which of the possibilities approaches the optimum technology.

This chapter is concerned initially with general considerations of the source − fibre combination, and describes the principal properties of a range of sources. Detailed source requirements which vary considerably for different sensors, and which often differ significantly from the requirements for PTT systems, are discussed in later chapters with descriptions of specific sensor or signal processing systems.

There are numerous detailed accounts of the principles and properties of optical sources. Kressel's book [3.1] is an anthology of individual articles on semiconductor devices for optical communications systems, while Yariv [3.2, 3.3] covers basic principles of optical sources and discusses in detail a number of examples. Thompson [3.4] presents an authoritative account of semiconductor laser diodes, and the optical communications anthologies [3.5, 3.6, 3.7] contain at least a chapter on lasers and LEDs. Light-emitting diodes have been covered in Bergh and Dean's book [3.8], though the emphasis here is more on display devices.

3.2 Source properties

It is convenient to specify an optical source in terms of four sets of properties, each to a large extent independent of the others. The groups are the geometrical characteristics of the emitted radiation, the optical properties of the radiation, the relationships between any applied electrical bias to the device and the consequent optical properties of the emitted radiation and environmental characteristics including thermal coefficients, long-term drift effects, aging mechanisms and so on.

3.2.1 Geometrical properties
As a general (but by no means universal) rule, it could be asserted that the basic criterion for a given source − fibre combination is that the maximum amount of light actually emerges from the remote end of the fibre. The amount which emerges is determined by a combination of the source wavelength (which determines the attenuation of the fibre) and the amount launched into the fibre.

The launch efficiency is directly related to both the fibre core area (or more strictly the area occupied by propagating modes in the fibre, which for single-mode fibres can significantly exceed the core area) and the fibre input numerical aperture. A linear optical nonabberating lens system will, in the ray optics limit, always conserve the product of area and solid angle of acceptance, as shown in Appendix 1. Thus the real criterion to determine the amount of power which may be launched from a given source into a given fibre is area × solid angle for the fibre and the power radiated per unit area per unit radiated solid angle. The latter parameter is known as the *radiance* of the source, and high-radiance sources are vital in any optical fibre system. Ideally, the area and numerical aperture of the source will exactly match that of the fibre, since any lens system will introduce abberations, which implies a reduction in the product of area and solid angle. In practice, lens structures are frequently used to optimise coupled power. The details of a lens structure depends on the spatial properties of the source, and are discussed for individual sources later in the chapter. For a single transverse mode system, Gaussian beam optics should be used. In principle any single transverse mode may be transformed to any other transverse mode.

It is useful, for reference purposes, to estimate the acceptance 'radiance' of a number of typical optical fibres, so that the required source radiance per milliwatt of launched power may be established. This then gives a measure of the source requirement for use in a typical fibre system − a typical launch power target is one milliwatt. These are shown in Table 3.1. This table very clearly indicates the relative levels of difficulty in launching light into the various types of optical fibres. Appreciation of the necessity for compatibility between source radiance and the fibre type is an essential factor in optical fibre system design.

3.2.2 Spectral properties of optical radiation
Specification of the spectrum of an optical source is not always so straightforward as may be imagined. In principle, the centre wavelength λ and the wavelength

Table 3.1   *Source radiances for various fibres*

Fibre type	NA	Guiding area cm^2 $\times 10^{-8}$	Area $\times$ solid angle cm^2 steradian $\times 10^{-8}$	Source radiance per milliwatt watts/(cm^2 steradian)
Single mode	0·15	50	0·28	3·6 $\times 10^5$
Multimode PTT	0·2	1960	19·6	5 $\times 10^3$
Typical PCS 250 micron core	0·5	4·9 $\times 10^4$	3070	33
Fibre bundle	0·5	7·8 $\times 10^5$	50000	2

spread $\delta\lambda$ should be adequate. However, there is often — especially in the case of laser sources — considerable fine structure within the range of $\delta\lambda$, and this can have important consequences on system performance. The noise level of the optical spectrum is also critical to analysing system performance, and again this is often far from straightforward to define or analyse, especially with laser sources.

The minimum noise level of an optical source is determined by shot noise, and in principle, optical systems should always operate with this as the limit (see also discussions on detectors in the next chapter). Shot noise is a direct consequence of the photon nature of light. If, on average, a source emits N photons in a given time interval, then there will be a variation of $N^{1/2}$ in the actual value emitted in successive intervals. This variation is equivalent to a noise component in the amplitude of the transmitted signal. If the photon energy is $h\nu$ and transmitted optical power P, then the number of photons emitted per second is $h\nu/P$. The photon noise (or shot noise) can thus be readily calculated (see for instance Reference 3.3).

In many optical sources, the actual noise level exceeds shot noise. This occurs for a number of reasons. The electrical supply which energises the source is often itself noisy, and this noise will be translated into optical noise through the optical/electrical transfer characteristic of the source. There are often resonance effects within the source — especially with lasers — which produce excess noise levels at specific frequencies. Plasma noise in helium neon (or other gas) lasers and the so-called 'self-pulsing' of semiconductor lasers are examples of this. There are also structural irregularities in the oscillating medium which give rise to inhomogeneous optical generation, and which may therefore produce instabilities which appear as noise on the output. Finally, there are variations in the optical phase length of the laser cavity which produce frequency variations in the output. These appear as phase or frequency noise (over and above the phase noise component of shot noise) and also as a (usually small) amplitude noise via the power/frequency relationship of the laser oscillator.

Most of these effects elude simple theoretical analysis on all but a superficial level. We shall restrict our discussions here to this very elementary treatment, though more detail may be obtained by reference to Thompson [3.4] for semiconductor lasers or Yariv [3.2] for many other lasing media. Perhaps paradoxically, most incandescent sources (which includes LEDs) have a noise level which is very close to shot noise. Intuitively, this may be explained by observing that there are no feedback loops in an incandescent source, and therefore no possible effects due to the frequency dependence of this feedback loop or due to unstable nonlinear effects within this loop.

The effects of noise in a signal processing or sensing system depend on both the type of noise — intensity or frequency variations — and on the design of the fibre optical system. In particular, there are often interferometric paths in a system whose path difference far exceeds the conventionally defined coherence length of the source. In such systems, minor phase noise in the source may give rise to significant increases in background noise level in the receiver. Specific cases of these phenomena are discussed later.

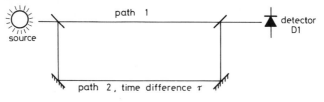

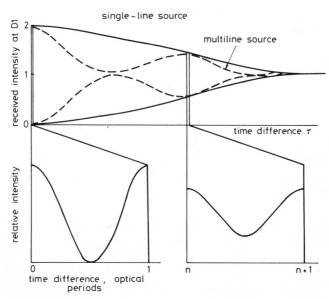

Fig. 3.1 *The relationship between fringe contrast, path difference and source coherence.*

Coherence is another critical parameter in matching the source to the system. Coherence is usually divided into two separate concepts, spatial coherence and temporal coherence. The first is a formal expression of how near a given source is to a point source, or alternatively how closely the emitted radiation from a source approaches a single spatial mode. Temporal coherence is the same concept in the frequency domain — that is, an expression of how closely a source approaches a single-frequency oscillator, or how closely its output is to a single temporal mode. Both the spatial and temporal coherence properties of the source affect application in interferometers. (For more detailed accounts of interference phenomena, see for instance the Open University [3.9], Hecht and Zajec [3.10], Francon [3.11] or Born and Wolf [3.12] — in roughly ascending order of mathematical thoroughness.)

The majority of optical fibre sensor and signal processing systems do involve some form of multipath operation between the source and the final receiver. There is thus the potential for interference effects, and it is therefore useful to consider the principles of the analysis of the effects of partially coherent sources in such a situation. The simplest case is that of combining radiation from a spatially coherent source (a single spatial mode) after half the generated power has been subjected to a different delay from the other half (see Fig. 3.1). If the delay corresponds to a time interval Δt, then the output at the receiver $f(\Delta t)$ is related to the source spectrum $F(\omega)$ by:

$$\int_{-\infty}^{\infty} F(\omega) F^*(\omega) d\omega = \int_{-\infty}^{\infty} f(t) f^*(t - \Delta t) d(t) \qquad (3.1)$$

That is, the output from the interferometer plots out the autocorrelation of the source spectrum. The coherence time of the source (often expressed as a coherence length, obtained by multiplying the coherence time by c, the velocity of light) is defined as that time difference at which the envelope of the autocorrelation function reaches some fraction — typically one-half, but sometimes $1/e$. Within this envelope are individual fringes as a result of normal interference effects (see Fig. 3.1). These fringes are related to the path difference Δl by:

$$I \propto \left\{ 1 + \cos\left(\frac{2\pi\Delta l}{\lambda}\right) C(\Delta l) \right\} \qquad (3.2)$$

where I is the output intensity from the interferometer, λ the wavelength of the light in the transmission medium and C the coherence function at path difference Δl.

The importance of these effects in optical fibre sensing and signal processing systems depends on the system involved. If the system is interferometric in nature, then the coherence function is a direct measure of the interference signal as a function of interferometer path difference. This signal is clearly a maximum at zero path difference, so that interferometric sensors are best used with zero path differences. At nonzero path differences, two effects occur. The first is that the maximum value (that is, the envelope) is decreased, but, and of equal importance, the minimum value *within the envelope* increases. The increase in the minimum value and the decrease in the maximum result in a decrease in the dynamic range of the output from the interferometer, and also reduce the scale factor (that is, the change in perceived optical intensity per unit change in relative phase in the interfero-meter). Additionally, the presence of frequency noise on the source raises the back-ground noise level as the path difference is increased, passing through a maximum until the autocorrelation function eventually decays to zero, when the noise perfor-mance returns to shot noise (or, specifically, the original source noise spectrum) and all interference effects cease.

In systems which are not interferometric but which do use sources which have partial temporal coherence, the situation is further complicated. It is usually simple to arrange that the optical processing mechanisms within the system occur outside the conventional coherence length of the source, but it is less simple to estimate the effects of source coherence on the noise level in the system. For example, modal noise [3.13, 3.14] occurs as a result of multipath interference effects in multimode fibres energised with coherent sources (often desirable because of their high radiance) in an incoherent (intensity modulated) system.

This straightforward simple relationship between the source spectrum and the interference term is complicated when multipath effects must be taken into account. The principle is simple; all that is required is to add the appropriate weightings and terms of the individual autocorrelation functions for each path. The practice is often mathematically complex (see also Appendix 3).

3.2.3 The electro-optic transfer characteristic
The electrical bias supplied to a source has a first-order effect on the optical output. The effect is multidimensional. Output level is usually increased with electrical drive, but device temperature is also usually increased with drive. For many electro-optic converters, this reduces the output somewhat when compared to a constant-temperature case. The optical frequency changes. For an LED this can be due to small bandgap changes as a function of temperature; with the laser this effect is compounded by thermal fluctuations in both cavity length and lasing medium refractive index, induced by both nonlinear effects in the lasing medium and the dependence of refractive index on electron concentration.

Thus, in general, both frequency and output intensity will be a function of elec-trical bias. Other effects also occur, especially in lasers. Excess noise contributions are strongly dependent on bias, for reasons which are not always directly obvious, and these are discussed in the context of individual sources.

The linearity of any relationship which exists between light output level and electrical bias is central to any discussion on the modulation of optical sources.

3.2.4 Environmental characteristics
With the exception of some semiconductor sources, most optical sources exhibit — to electronics industry standards — lamentably short mean times to failure, usually in the region of a few thousand hours. There is also frequent degra-dation in output with aging and distinctly strong thermal coefficients. The failure mechanisms are, in many cases, little understood, though some have been charac-terised for activation energies etc. Often the simplest approach in practice is to

incorporate a maintenance step to renew the source. This is usually unneccesary with semiconductor sources, but depends, even here, on the details of the device technology, and thorough investigations are required for each individual source under consideration.

3.2.5 Source parameters – discussion

Relating an optical source to a particular system requirement involves analysis of numerous interacting parameters. The principal criterion is that sufficient light of the appropriate quality reaches the detector to ensure adequate signal to noise ratio. The noise contributions include shot noise in the light beam and the detection process, detector thermal noise, noise figures in successive amplification stages and excess noise contributions from the source itself. Spurious random modulation mechanisms within the system also occur. In particular, there is always the often neglected possibility of interference effects due to inevitable path differences introduced by multiple paths, delay lines etc., which always add to the receiver noise for source with any degree of coherence.

Then the interaction between the various source parameters – radiance, spectral characteristics and electro-optic transfer characteristics – plays a central role in the design of sensor and signal processing systems. Aging effects are equally important, though in many cases the relevant aging process is still ill defined for these applications. In the following sections of this chapter, we describe the properties of a range of readily available and commonly used optical sources for the systems under consideration. An analysis of the match between the source characteristics and the system requirements will be left to the discussion of specific systems.

3.3 Optical sources for fibre sensor and signal processing systems

3.3.1 Introduction: classification of sources

Optical sources can be placed into two very broad groups – coherent and incoherent. The former category consists largely of laser devices, which exhibit a high degree of spatial coherence, that is, they are high-radiance sources, with varying degrees of temporal coherence. Lasers consist essentially of a resonant optical cavity filled with a medium which exhibits optical gain. The gain has a finite bandwidth, determined by the range of energy level differences which are available; usually the cavity will have numerous resonances in this gain bandwidth, and typically the laser will oscillate in several frequencies simultaneously – that is, in several modes. The optical frequency resonances within the lasing cavity are quite complex, consisting of numerous permitted field distributions both longitudinally and transversely. The transverse field distributions introduce higher orders into the spatial coherence of the laser, and modify the directional characteristics of the emerging beam. Most but by no means all lasers are controlled to oscillate in only one transverse mode. Each mode then corresponds to a different temporal frequency, so that the temporal coherence is determined by both the longitudinal and transverse modal fields, and is different for each combination. This is shown diagrammatically in Fig. 3.2. In principle, it is a straightforward spatial filtering problem to remove all except one selected transverse mode, but it is considerably more difficult to filter out the various longitudinal modes corresponding to each transverse mode. The various longitudinal modes are usually not mutually phase locked, so that the output is a sum of the total mode spectrum varying in spatial distribution corresponding to the relative phases of each mode. In mode locked lasers the relative phases are locked, producing regular pulses of optical power (see Reference 3.2).

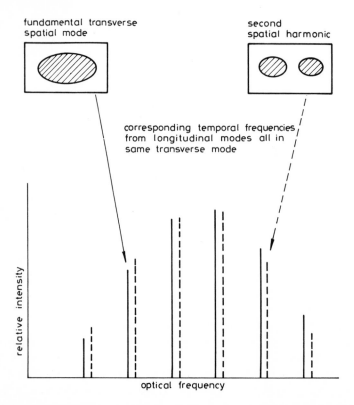

Fig. 3.2 *Illustrating different longitudinal mode spectra corresponding to two different transverse modes*

The coherence length of a laser source depends on its mode structure and on the individual mode linewidths. The coherence function is defined by the visibility of fringes produced in an equally split two-arm interferometer. The visibility V is defined as:

$$V = \frac{I_{\max} - I_{\min}}{I_{\max} + I_{\min}} \tag{3.3}$$

where I_{max} is the maximum intensity of the fringe pattern at a particular path difference, and I_{min} is the minimum at the same (or strictly one half wavelength different) path difference. This visibility function may be readily shown to be the autocorrelation function of the laser output, which is related to the spectrum by Eqn. 3.1.

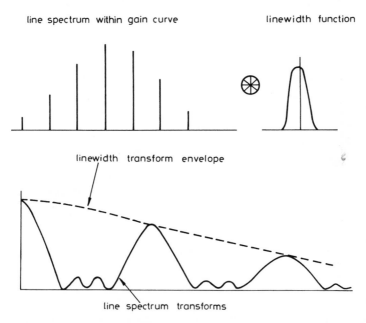

line spectrum within gain curve   linewidth function

linewidth transform envelope

line spectrum transforms

Fig. 3.3 *Source line spectrum represented as a line spectrum convolved with a linewidth function, and the corresponding coherence function. Typically, the linewidth will increase with decrease in mode power, but the general behaviour is little altered*

The coherence function determines the path difference over which an interferometric system may operate. A laser output may be viewed as a line spectrum convolved with a linewidth function (see Fig. 3.3). The form of the coherence function is then determined by the spacing of the line spectrum. If these are equally spaced, then the coherence function will be regularly repeating within an envelope defined by the linewidth (see Fig. 3.3). If the lines are not regularly spaced, then the coherence function rapidly falls in a time of the order of the reciprocal of the deviation from constancy of the frequency difference. Notice that the coherence function does not fall to zero until the envelope determined by the linewidth falls to below a level determined by other noise levels in the source spectrum. As an example, the Lorenzian lineshape is given by:

$$P(\nu) = \frac{\Delta\nu/2\pi}{(\nu - \nu_0)^2 + (\Delta\nu/2)^2} \qquad (3.4)$$

The reciprocal of the full width half maximum, $\Delta\nu$, gives the coherence time (and the coherence length when multiplied by c, the velocity of light in the propagating medium). However, interference effects will exceed the intrinsic noise level of the source for some considerable further distance. As an example, for 160 dB peak source level to shot noise level, interference effects will continue to be important as a noise source until the coherence function drops below the noise level. This will occur when the Lorenzian linewidth function is about 10^{-8}, corresponding to approximately 4000 times the linewidth. Thus, significant coherence effects can occur at distances of several thousand times the conventionally determined coherence length, and these will manifest themselves as an increase in background noise. Paradoxically, this effect is important primarily in nominally incoherent systems.

Incoherent sources represent an opposite extreme. These are omnidirection in their radiation pattern, or Lambertian if they are surface emitters. A Lambertian source has a radiation pattern given by:

$$I(\theta) = I_0 \cos \theta \tag{3.5}$$

where θ is the angle between the direction considered and the normal to the radiating surface.

Incoherent sources may be thought of as optical noise generators, the power output of which is the variance of the noise waveform generated. In treating the details of the output from an incoherent source, it is important to consider the fact that the constant output from, say, a light-emitting diode is really the envelope function of a band-limited random noise source with a coherence time of the order of the reciprocal of the emitting linewidth. Processes which occur on a shorter timescale than this coherence time will respond to the source variations, and should therefore be treated statistically. For instance, for a light-emitting diode the coherence time is typically of the order of 0·1 ps, and many atomic processes are faster than this. Processes which occur on a longer timescale may be considered as responding to the average intensity of the noise source. Only in cases in which the process considered responds linearly to the amplitude of the light vector will the two calculations produce identical results; hence time constants are a central issue in the interaction between a light source and a physical mechanism.

The low-frequency noise spectrum of an incoherent source is shot noise, unless there are irregularities in the power supply to the source. There is no optical feedback, so there are no cavity resonances to contribute to the noise spectrum. This is a very important distinction between coherent and incoherent sources.

A wide variety of optical sources is used in sensors. Accordingly, the remainder of this section is devoted to brief discussions of incandescent sources, gas lasers, solid state (crystal) lasers and semiconductor sources.

3.2.2 Incandescent sources
A light bulb converts essentially all the electrical energy input to it into radiation. The filament radiates as a black body (or approximately so) but over a very broad band of wavelengths. For black bodies, the Stefan-Boltzmann Law predicts that at 2000 K (an average filament temperature) the total radiated power will be in the

region of $80 \, \text{W/cm}^2$. This gives an equivalent radiance – since the radiation will generally spread over a full 4π steradians – of about $6 \, \text{W/(cm}^2 \, \text{steradian)}$. Recall that this is total electromagnetic radiated energy. The percentage of this energy in the band of interest – typically the visible and near infra-red wavelength region – will be small, so that the useful radiance at 200 K will be in the order of $0 \cdot 1 \, \text{W/(cm}^2 \, \text{steradian)}$.

We may observe that the Stephan-Boltzmann Law predicts an increase in total radiance proportional to T^4 and the Wien Displacement Law indicates that the peak wavelength of the emitted spectrum is inversely proportional to T, so that the power in the visible spectrum increases roughly as T^5 (T is the absolute temperature of the radiating body). Hence, the use of a slightly higher filament temperature, say of 3000 K, will increase the useful radiance to the order of $1 \, \text{W/(cm}^2 \, \text{steradian)}$, but with a reliability penalty. Tungsten lamps, in particular, are often used as sources for certain classes of sensors. These run at high temperatures, and consequently have lamentably low lifetimes (a few hundred hours), but useful optical power may be launched into a fibre to a remote environment. Tungsten lamps are also an essential part of crystal lasers, providing the pump power. Here they are again run at very high temperatures with low lifetimes.

If we refer back to Table 3.1, we see that useful power can be launched from an incandescent source to a fibre bundle, and possibly into a large-core step index fibre. If other fibres are required in the system then a different source must be found.

3.3.3 Gas laser sources

Gas lasers are useful general purpose sources for applications requiring a high degree of coherence. The most frequently used gas laser sources are HeNe, operating at $0 \cdot 633$ microns or $1 \cdot 15$ microns, carbon dioxide, operating at $10 \cdot 6$ microns, and argon ion lasers operating at $0 \cdot 516$ microns.

The HeNe laser is particularly useful as an inexpensive, low-power ($0 \cdot 1$ to 100 mW) highly coherent source. The longitudinal mode spectrum of an HeNe laser is determined by the gain curve of the lasing medium and the resonant frequencies of the cavity. The modes are very nearly regularly spaced in frequency (provided that they all have the same transverse modal structure). There is, however, a small and thermally varying dispersion effect in the gain medium, resulting in occasional low-frequency osciallations in the laser output during warm-up. Under stable thermal conditions, the laser usually has a coherence function typical of the ideal regularly spaced spectrum.

The gain backwidth of a HeNe laser, like all gas lasers, is determined in most circumstances by Doppler broadening of the very narrow emission line between the relevant atomic energy levels. The Doppler broadening is a direct consequence of the Doppler shift produced by atomic motion, and is readily calculated from the temperature of the lasing medium and the atomic mass. For a HeNe laser, this typically turns out to be a gain bandwidth of approximately $1 \cdot 5 \, \text{GHz}$. The cavity length determines the mode spacing, so that it is a straightforward procedure to determine the number of oscillating modes (see Fig. 3.4).

Single-longitudinal-mode operation is fairly readily achieved in a HeNe laser, one technique being to reduce the cavity length so that only one resonance occurs in the lasing bandwidth. This single mode can have a very narrow linewidth, and as low as 1 kHz has been achieved (corresponding to a coherence length in air of 300 km). A few kilohertz linewidth is typical of readily obtainable single-mode HeNe lasers.

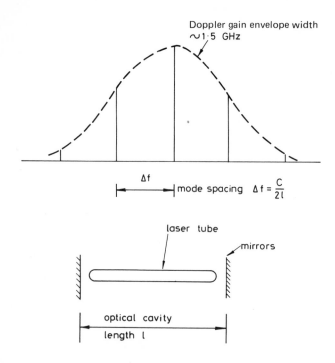

Fig. 3.4 *Relationships between mode spacing, in frequency, cavity length and the gain curve of the lasing medium*

All gas lasers are pumped (that is, the higher energy levels of the atoms in the lasing medium are filled before lasing) using some form of electrical discharge. This discharge ionises the atoms in the laser cavity, forming a plasma. Plasmas are known to exhibit plasma resonance, related to the plasma density, and this is often excited in the laser. The resonant frequency for the low densities in an HeNe laser is typically in the low megahertz region, and so there is usually a plasma noise peak at a frequency corresponding to the plasma resonance. This is inherent in all HeNe gas tube lasers. Its intensity varies with power supply levels, aging and temperature. It is usually preferable to avoid the relevant frequency band in designing a system utilising such a laser.

Throughout the remainder of the frequency range, the HeNe laser is usually a very quiet source, with noise levels very close to the shot noise limit. There are

$1/f$ effects at low frequencies, and multimode second-order beats occasionally fall in the usable frequency band when three or more modes are oscillating. The first-order beat frequency is typically in the hundreds of megahertz range, and though important for some high-speed operations, for instance in signal processing, has little influence on most sensor requirements.

The HeNe laser is also a very convenient very-high-radiance source. A 1 mW output typically originates from an area of 1 mm diameter and emerges with a divergence of about 1 m rad. Simple arithmetic quickly shows that this corresponds to a radiance of approximately 10^8 W/(cm^2 steradian) in a single transverse mode. High efficiency coupling to single-mode fibre optics is thus relatively straightforward.

The HeNe gas laser is the most important gas laser in the context of this book. Argon ion lasers produce very high powers (watts) from a similar aperture, and thus with very high radiance. They are expensive, inefficient and inconvenient to use, but are useful for the investigation of nonlinear effects and for holographic processes where large areas are to be coherently illuminated. Carbon dioxide lasers are higher powered still (hundreds of watts) but operate in the far infra-red, and are therefore used as cutting tools and as the source for lidar (optical radar) systems for probing atmospheric composition and similar applications.

3.3.4 Solid state (crystal) lasers

Lasing action was first observed in a solid state laser — a ruby crystal. Ruby lasers are now rarely used, but there are a number of solid state sources (not to be confused with semiconductor sources) which are important. The most common is the neodinium YAG system in which Nd^{3+} ions in the YAG crystal are the lasing medium. Typically this will operate at a wavelength of $1 \cdot 064$ microns, though other lines are occasionally used. The gain bandwidth is approximately 10^{11} Hz; this is considerably more than the HeNe laser, owing to the effects of the crystal structure spreading the energy levels in the lasing medium.

The crystal laser is pumped using some form of flashlamp source, and for CW operation this will typically be a high-output tungsten lamp. The lamp is placed close to the lasing crystal, and since the lamp rating will be typically in the kilowatt range, water cooling is always required. Even so, flashlamp lifetimes remain in the hundreds of hours. The laser output is related to the pump input power and also to the temperature of the crystal. It is thus preferable that the lamp be DC powered for a stable source, and it is also necessary to stabilise the cooling systems if high spectral stability is required. The Nd:YAG laser is principally used as a cutting tool, so that spectral stability is not at all important. In sensing and signal processing systems the device can be useful as a high-power, high-coherence source, though spectral stability is a problem. The lasing wavelength also corresponds to a low detection efficiency in silicon photodiodes, so that germanium or III-V compound detectors are required. The available power densities are very high. The lowest CW output Nd:YAG laser is rated at 100 mW, and tens of watts CW can be achieved. These correspond to radiances up to more than 10^9 W/(cm^2 steradian).

3.3.5 *Semiconductor light sources*

Semiconductor light sources are probably the most important of those used in sensor and signal processing systems. There are obvious advantages of small physical size, high reliability, adequate radiance levels and simple power supply requirements. There are also some uncertain areas. In particular, the coherence properties of semiconductor sources are sometimes difficult to interpret in the context of a particular application, and unusual effects are often observed.

Semiconductor sources differ in other, perhaps more fundamental, respects from gas and solid state devices. The same basic device (with considerable structural differences) acts as both an incoherent source (the LED) and as a laser (the injection laser diode), and similar physical mechanisms are involved in each case. The width of the gain bandwidth is higher than in any other medium, essentially because the photon emission takes place as a result of electron motion between two bands of energy levels. The gain curve in a semiconductor laser typically extends over 10^{12} Hz.

The semiconductor laser differs from other laser sources in several important parameters. In particular, the active medium also forms the structure of the resonant cavity, and the termination of the active medium is also the end of the cavity. In all other lasers the cavity mirrors are separate from the active medium. The optical energy densities in semiconductor lasers are extremely high, and this leads to significant nonlinear effects which influence the device characteristics.

Much has been written on the subject of semiconductor lasers and light-emitting diodes; see, for instance, References 3.1 to 3.9 where the details of the device operating mechanisms are very adequately covered. Here we simply recap a number of basic points which are common to all semiconductor radiating devices. The wavelength of the emitted radiation is determined by the energy gap of the semiconductor and the width of the gain curve by the density of states function. The relationship is simply that:

$$\lambda \sim hc/E_g q \tag{3.6}$$

(for E_g in electron volts) and this is conveniently expressed as $\lambda = 1\cdot24/E_g$. The lasing bandwidth peaks at about 20–30 mV less than the bandgap energy. For incident photon energies less than the bandgap energy, a semiconductor is virtually transparent; for energies above, the material is essentially opaque. The lasing region is on the absorption band edge and so, since it is a fundamental property of absorption bands that the sharp increase in loss corresponds to a rapid change in refractive index over the same wavelength range, the semiconductor is a highly dispersive medium around the lasing wavelength. This can have important effects on the radiation properties. This feature also implies that any radiation leaving the active region of the device and passing through the semiconductor medium will probably be strongly absorbed within the intervening material. This leads to the frequent use of heterostructures in semiconductor sources. A heterostructure is one in which the materials are varied from layer to layer within the device. The materials are carefully chosen so that the lattice constants are fixed, thus ensuring continuation of the crystal planes from one layer to the next, while the bandgaps are varied to cater for desired variations in the optical properties from one layer to an adjacent one.

The remainder of this discussion on semiconductor sources is devoted only to the user aspects of the devices. It will become apparent that there are a number of practical features which must be taken into account when using semiconductor sources, and also that, especially for semiconductor lasers, some care is required in specifying a source to meet the system requirements.

3.3.5.1 Light-emitting diodes: Light-emitting diodes used in fibre systems differ considerably from the familiar display devices. The emphasis is on high radiance for optimum coupling to the fibre, and with this comes high speed, high reliability and choice of emission wavelengths to coincide with windows in the fibre attenuation spectrum. Both homostructure (that is single-material devices) and heterostructure diodes have been evaluated. The latter generate higher optical power. The overall designs fall into two broad groups, surface emitters and edge emitters. Some of the latter fall into a category known as super-radiant devices.

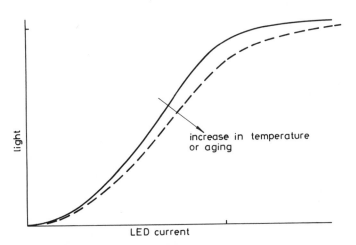

Fig. 3.5   *Light/current curve for a light-emitting diode*

There are a number of features common to all LED devices. The light/current curve follows the behaviour shown in Fig. 3.5. The variation of light output with forward bias current is close to linear over a wide range – perhaps 40 dB – but there is a pronounced variation in the position of this curve as the device temperature is varied. This has important effects on the design of source systems. Some thermal feedback must be incorporated, whether in the form of a temperature-sensitive supply or through monitoring via a photodiode in the transmitter assembly.

A light-emitting diode may be considered to be a band-limited Gaussian noise source, extending over an optical frequency range dictated by the bandwidth of the energy level differences between which emission takes place. This is typically a few times 10^{12} Hz. Within the modulation frequency (baseband) region, this implies that the radiation is effectively shot noise limited unless spurious noise is transmitted from the bias supply network. The coherence length of a LED source is in the range of a few microns.

A light-emitting diode may be directly amplitude modulated, simply by varying the bias current, and the process may be used as an analogue modulation technique, but with limited linearity unless feedback is used. The achievable modulation rate depends critically on the structure. Surface emitters can be modulated up to a few hundred megahertz, but modulation rates in excess of 1 GHz have been reported for more compact edge-emitting device.

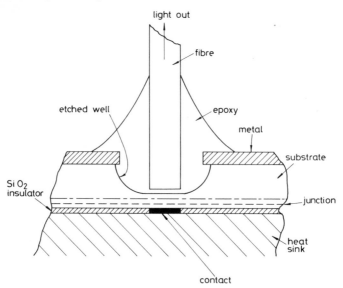

Fig. 3.6 *Section through the Burrus diode structure*

A surface-emitting diode as used in optical fibre systems is shown in cross-section in Fig. 3.6. This is often referred to as the 'Burrus diode'. These devices are 'pigtailed' to a fibre, and are excellent sources for multimode optical fibre systems. They are capable of generating optical power of the order of 1 mW from a surface area of 50 microns diameter in a solid angle of 2π steradian, corresponding to a radiance of 25 W/(cm² steradian). Surface emitters with a radiance up to a few hundred W/(cm² steradian) appear to be feasible. Coupling a surface emitter to an optical fibre is inherently an inefficient process, simply because the generated power occupies too great a solid angle, and less than 10% typically finds its way into a PTT fibre. Clearly much here depends on the fibre type, the optical structure between the emitting surface and the fibre end, and the details of the design of the emitter itself. Much higher efficiencies may be achieved by using larger fibres. The radiance of a surface emitter is so low that it is impractical to consider using a surface emitter with a monomode optical fibre system.

Edge emitters are somewhat of a halfway house between a semiconductor laser and an LED. They have much higher spatial coherence; in fact the radiating aperture (see Fig. 3.7) is typically similar in size to that in a laser. However, the active

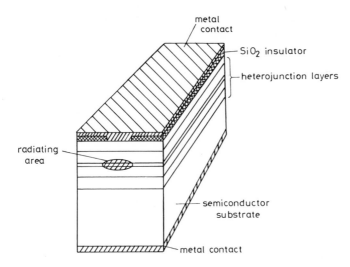

Fig. 3.7   *Schematic representation of an edge-emitting LED*

region of the device is designed to inhibit feedback, and so prevent the onset of lasing action. Thus the light output is temporally incoherent. The total amount of power generated by an edge emitter is smaller than that from a surface emitter, but the radiance is significantly higher. At 1 mW, an edge emitter may have a radiance of over 10^3 W/(cm^2 steradian). The polar diagram of the source is now no longer completely Lambertian (see Fig. 3.8) but is astigmatic, owing to the narrow radiating surface in the plane of the semiconductor junction. There is also some limited optical feedback, resulting in limited linewidth narrowing when compared to the surface emitter. Optical linewidths of a few nanometres are common, corresponding to bandwidths of 10^{11} Hz and coherence lengths of tens of microns. The high radiance of the edge emitter facilitates the coupling of significant power to PTT multimode fibres and useful power to single-mode fibres. The optical waveform generated by the device may still be considered as a noise source and the noise levels at baseband are close to shot noise, again neglecting power supply irregularities.

3.3.5.2 Semiconductor laser diodes: Semiconductor laser diodes are very versatile, high-power (up to tens of milliwatts) sources of optical energy. The radiation from a laser diode does however have a number of curious behavioural characteristics. A sketch of a typical laser diode is shown in Fig. 3.9. The semiconductor crystal is a heterostructure to encourage guiding of the light in the very narrow active region, defining the vertical limits of the cavity. The transverse limits of the cavity are defined by the passage of current through the narrow stripe contact and/or by the refractive index profile of the semiconductor structure. The current flow produces a slight refractive index change. The longitudinal limits of the cavity are defined by the cleaved facets of the semiconductor crystal.

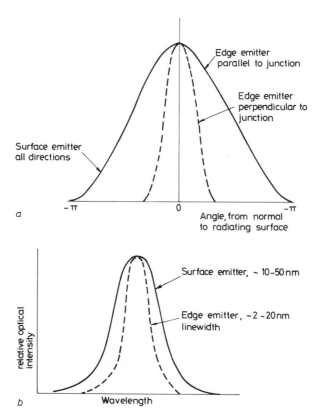

Fig. 3.8   *Radiation characteristics from surface and edge emitting diodes (a) directional and (b) wavelength spectrum*

The field distributions of the fundamental and first higher transverse mode of the cavity are shown in Fig. 3.10. These field distributions give some clues to some of the behavioural characteristics of the device. Note that the fundamental transverse mode exhibits a high degree of symmetry about the central axis of the cavity, and then recall that the transverse guiding mechanism is largely though electric current density variations. Any slight deviations from the required symmetry in the current flow can produce changes in the cavity symmetry, and thus promote the development of higher-order transverse modes. Note also that the field extends into the material surrounding the active region. The interaction between an optical field and a semiconductor material when the field is at a frequency corresponding closely to the bandgap energy and when the field is also crossing a junction where the bandgap is a rapidly varying function of distance is, to say the least, complex. It can be solved, but in practice varies considerably from laser to laser. The energy densities in the field within the cavity are very high, so that nonlinearities are also important. In particular, the edge region often acts as a saturable absorber — that is, it is lossy up to some field level, above which the loss decreases significantly.

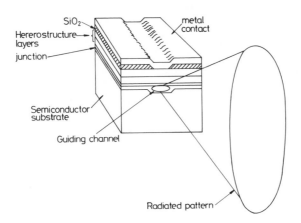

Fig. 3.9   *Diagram of an index guided semiconductor laser structure*

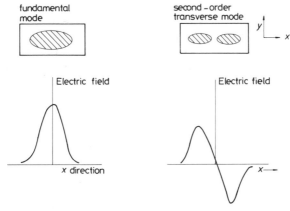

Fig. 3.10   *Fundamental and second-order transverse modes in semiconductor lasers — intensity distributions and electric fields*

This brief and qualitative look at the basic physics of a semiconductor laser will indicate that there are numerous possible feedback loops within the structure of the device itself. Interactions between variations in current density across the junction, electric field distributions, particularly in the transverse direction, and nonlinearities both in the lasing cavity and in the material surrounding the cavity lead to the possibility — and the fact — of complex characteristics.

One final aspect of the basic device characteristics, and a total contrast with the lasing devices considered in previous sections of this chapter, is in the mirrors which define the ends of the cavity. In gas and solid state lasers, very high reflectivity mirrors are used, and these are completely separate from the active medium. In a semiconductor laser, the cavity ends are defined simply by Fresnel reflection at the end face (that is, reflection due to the refractive index difference between the cavity and the surrounding air). The reflectivity of this interface is only about 30%. Consequently, reflections from other interfaces − for instance, the 4% reflected at a glass−air interface at the end of a fibre − can return to the active region, producing effectively an extension of the laser cavity. Often the external reflection back to the laser is environmentally sensitive, and so the laser output can vary − sometimes significantly − with variations in what may be termed the optical load. The same effect is observed with HeNe lasers, but at a lower level.

This provides a background for discussion of the operational characteristics of the semiconductor laser diode. Most semiconductor lasers emit power levels of the order of 10 mW and, most commonly, at a wavelength of 850−900 nm. This wavelength is particularly suited to the use of silicon photodiodes, since it coincides almost exactly with the peak response of that material. There is currently much interest in fibre systems operating at 1·3 and 1·5 micron wavelengths, and laser sources are also becoming available for use at these wavelengths. These longer wavelengths may be useful for sensor and signal processing applications, not because of the improved attenuation and dispersion characteristics but primarily because of the availability of fibre components for use at these wavelengths.

The optical power radiated from a semiconductor laser (see Fig. 3.9) leaves a rectangular aperture at the end of the lasing cavity. The aperture is less than 0·5 microns in depth and from 5 to 20 microns in width. The radiation is coherent across this aperture (contrast the edge-emitting diode) and so will have a far field pattern related to the field at the end of the cavity via a Fourier transform. For the fundamental transverse mode, the far field from the laser will be in the form of the astigmatic cone shown in the figure. The half angles will cover a range from 30 to 60 degrees vertically and 5 to 20 degrees horizontally. The radiance of these devices may be readily estimated; 10 mW originating from 10 microns by 0·5 microns into a cone of $45°$ by $5°$ half angles gives a radiance of the order of 10^8 W/(cm^2 steradian). The angular distribution of the power along the two directions will, for the fundamental transverse mode, be of the form indicated in Fig. 3.11*a*. The second-order transverse mode can, and frequently does, occur, in which the field at the end of the cavity is shown in Fig. 3.11*b*. Here the far field will occupy a similar total cone angle, but with a radically different spatial distribution.

The more detailed properties of semiconductor lasers vary over a quite bewildering range of possibilities. We need to consider the principal features of the light/current curve, the noise performance and the coherence performance of the devices.

The light/current (L/I) curve is of the general form shown in Fig. 3.12. There is a low-level region, an abrupt transition (the threshold current at which lasing action

takes over from spontaneous emission) and a rapidly varying region beyond the
threshold current. The threshold current is then a key parameter in defining the
performance of the laser diode. The value of I_{th} increases with temperature and

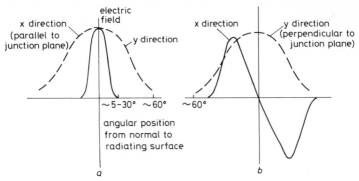

Fig. 3.11 *Far field radiation patterns for (d) fundamental and (b) second-order transverse
modes*

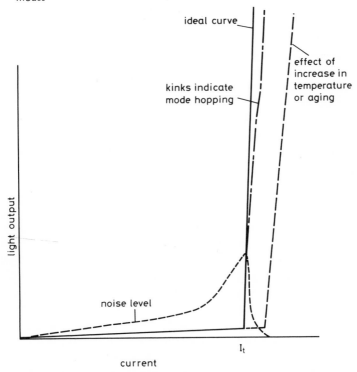

Fig. 3.12 *Light/current curves for a semiconductor laser. I_t is threshold current.*

increases with aging of the diode. There is thus an automatic prerequisite for some
form of output monitoring in the laser diode. This will typically be a local photodiode providing a reference signal to adjust the bias current accordingly.

In a well behaved laser diode, the L/I curve will be smooth and approximately linear above threshold over a wide range – up to the onset of damage within the lasing structure. Damage occurs at high light output levels. This smooth curve indicates that both the transverse and longitudinally mode spectra are stable. This implies that either the device is oscillating in a single stable longitudinal and transverse mode, or it is maintaining a single transverse mode but has a multiple-line longitudinal spectrum. In the latter case the longitudinal spectrum will be changing in detail, resulting in partition noise, but will maintain a constant overall level. The presence of kinks in the L/I curve is a reliable indicator of changes in the longitudinal and/or transverse mode spectrum of the device. Each mode will have a slightly different lasing threshold, and will require the presence of the correct symmetry to ensure excitation of the mode. The current distribution within the cavity varies as the current level varies, owing to the appearance of local 'hot spots' etc., so that the modal preference also varies. Therefore there are abrupt changes in the mode spectrum, and hence output level, of the device. These kinks are commonly observed in the L/I curves of a semiconductor laser. In some applications they are of no consequence. In others, the change in the modal structure can have important implications.

The longitudinal mode structure of the laser diode typically follows a pattern as shown in Fig. 3.13. Here the development of the lasing action from the spontaneous emission case is shown for both a multilongitudinal-mode diode and the single-mode diode. The mode structure in the multimode case does vary with drive current, temperature, aging etc., but provided that there are an adequate number of modes in the spectrum – say greater than ten – the device will behave as if the mode spectrum is practically constant. As one mode disappears, another will take its place, and the total output will vary smoothly. Highly multimoded lasers are particularly useful for applications involving analogue modulation of the source.

The lasing cavity is very dispersive at the lasing wavelength, so that the modes in a multimode laser are not equally spaced in the frequency domain. The coherence function is then the transform of a series of frequencies that are not harmonically related, and so does not follow the long repeating format typical of the helium neon laser. The individual linewidths in the spectrum may be in the region of a few megahertz or less, but the overall coherence function will extend over a matter of millimetres. This is not to preclude the possibility of coherent interference effects at significantly greater path differences. In fact, some interference can occur for temporal path differences up to the reciprocal of the linewidth. In Fig. 3.14 the coherence function, is plotted on a log scale, for a five-moded laser with dispersion between the modes, with and without a finite linewidth effect. With an assumed linewidth of 1 MHz, it can be seen that some measureable (that is, greater than shot noise) interference effects occur for path differences up to several hundred metres, even though the 'coherence length' defined by the 3 dB point is only a few millimetres,

Noise in semiconductor lasers appears from all manner of sources. There is the multimode interference noise alluded to here. There is what may be termed 'relaxation noise' which occurs due to interactions within the lasing cavity with time constants of the order of nanoseconds. This varies considerably between devices, though some models for a general noise source have been proposed [3.15, 3.16]. Then there is spontaneous emission noise. Before the onset of lasing action, the semiconductor laser diode acts as a low-power light-emitting diode. At threshold this light is passed through a tuned amplifier, and appears as the lines in the laser output spectrum. Just before the onset of lasing, the noise level from the source passes through a strong peak. Similarly, mode hopping effects, corresponding to kinks in the *L/I* curve, also coincide with pronounced increases in noise level.

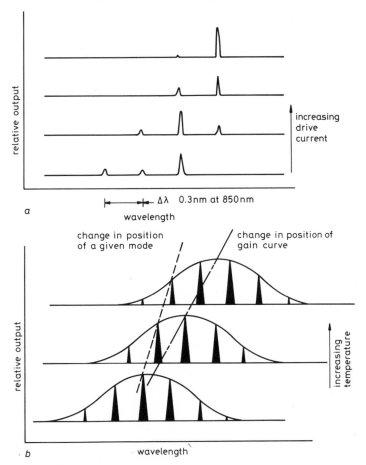

Fig. 3.13   *(a) Change in laser spectrum with current for a single-mode semiconductor laser at constant temperature, and (b) changes in multimode laser spectrum at constant current with increasing temperature*

Associated with the noise behaviour is the so-called self-pulsing behaviour of some laser diodes. Self-sustained oscillations in the power output — thought to be caused by nonlinear relaxation effects in the cavity region — cause noise peaks at frequencies in the 1 GHz frequency range. It is conceivable that much higher frequency self-pulsing also occurs. This particular type of behaviour is difficult to relate in a precise way with device physics, though some structures appear to be less suscep-

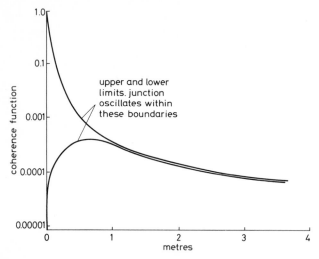

Fig. 3.14   *The coherence function for a five-moded semiconductor laser including the effects of finite linewidth and intermode second-order dispersion. Five modes are of equal amplitude, spaced 100 GHz apart, with 10 MHz second-order dispersion, 1 MHz linewidth*

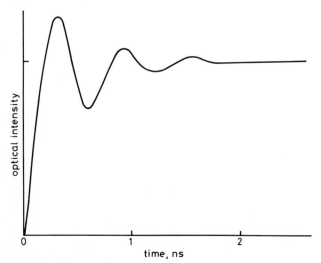

Fig. 3.15   *Typical transient ringing in semiconductor lasers*

tible to the effect. Various means of forming a more precisely defined lasing region appear to help.

This self-pulsing effect has important consequences for the transient response of

Table 3.2a   *Light sources — optical and spatial properties*

Source	Centre wavelength	Total spectral bandwidth	Number of oscillating lines	Individual linewidths	Spatial properties	
					Polar diagram	Transverse spatial mode spectrum
Incandescent	Usually black body radiators,[5] but some lamps (for instance; mercury lamps) have line spectra				Omnidirection or Lambertian	
Solid state laser Nd YAG	1·064 microns	10^{11} Hz (0·3 nm)	Several[4]	Usually megahertz	Slowly diverging beam (Gaussian)	Few, single transverse mode available
Gas laser HeNe[2]	0·6238 microns	10^9 Hz (0·0013 nm)	One or several[4]	Typically 10^4–10^5 Hz	slowly diverging Gaussian beam	Single transverse mode usual
Semiconductor (GaAlAs) laser[3]	820–900 nm	10^{12} Hz (2·4 nm)	One or several[4]	Typically 10^5–10^6 Hz	Elliptical rapidly diverging Gaussian beam	Single transverse mode usual, but higher orders often observed
Semiconductor (GaAlAs) LED	800–900 nm	20–30 nm (~10^{13} Hz)	Broadband source[5]	N/A	Lambertian	No mode structure
Semiconductor (GaAlAs) Superluminescent diode	800–900 nm	~5 nm (2 × 10^{12} Hz)	Broadband source[5]	N/A	Lambertian in one direction, slightly focused in the other	Sometimes single spatial mode available. Depends on structure

[1] Nd YAG also has other lines available, and power outputs are high enough to use optical frequency doublers for precisely defined wavelengths in the visible (400–700 nm) range

[2] Other HeNe lines include an especially useful one at 1·115 microns. Power levels are lower and the 633 nm line is more usual

[3] Other semiconductor laser and LED sources are available for the 1·3 and 1·55 communications bands

[4] Lines are spaced in frequency as dictated by external cavity length

[5] Broadband sources may be considered as band-limited white noise sources

Table 3.2b   Light sources — electro-optic properties and comments

Source	Properties	Comments
Incandescent lamps	Light output a strong function of electrical input power. Useful in applications where long-term stability not required unless feedback monitoring incorporated. Power can be high (even watts) but incoherent, spatially and temporally.	Spatial coherence only suitable for use with high-NA fibres. Useful as UV short-range. Lifetime ~ 1000 hours typical. Noise is shot noise plus current drive effects
Solid state lasers Nd YAG	Laser optically pumped — so output depends on optical input. Water cooling implies relatively unstable lamp filament temperature unless feedback is used; typically, output is intensity unstable. Frequency spectrum also varies with crystal temperature, and hence with minor fluctuations in cooling system. Power 100 mW CW upwards	Pump lamp life typically a few hundred hours. Very difficult to obtain single longitudinal mode. Spectrum drifts with coolant fluctuations and with lamp intensity fluctuations. Noise depends on both thermal and electrical variations in pump lamp
Gas laser HeNe	Output depends on drive current in electric arc. Can be thermally stabilized. Deteriorates with age. Power 1 to 100 mW typical	Plasma noise gives noise peak in the MHz region. Tube life typically few thousand hours
Semiconductor laser	Details of mode spectrum difficult to predict and depend on source structure and electrical drive. Can be made so that output power linearly proportional to current over threshold. Power to 50 mW CW. Can be intensity modulated to GHz	Complex noise resonance effect especially in the GHz region. Optical spectrum depends on both temperature and current. Good lifetime predictions
Semiconductor LED	Linear dependence of light power on current over reasonably wide range (approx 30 dB with correction circuits). Light/current curve lowers light output for a given current with both aging and temperature increase. Power typically less than 1 mW	Useful only with multimode fibres because of spatial properties of output. Good reliability. Shot noise
Superluminescent diode	Light/current curve similar to LED, but spatial coherence increased (perhaps to single mode) and linearity and stability similar to LED. Power typically 1 to 3 mW	Some structures spatially compatible with single-mode fibres. Relatively new structure, and lifetime looks good, but has to be proven. Good as a temporally incoherent, spatially coherent source. Shot noise

the device when the bias current is changed on a subnanosecond timescale. The light output rings, often for several cycles (see Fig. 3.15), before reaching the stable value. Again the effect varies considerably from one device to the next.

Semiconductor lasers are available in multimode narrow stripe configurations, as single-mode devices, and as devices with an unspecified output spectrum. They are also available with self-contained power supply units, with bare diodes and with diodes pigtailed to both single and multimode optical fibres. They are excellent optical sources for fibre systems, but their coherence properties, described very much in outline here, can produce apparently anomalous system behaviour. The device parameters are being continually improved, and the understanding of the systems behaviour is continually increasing. The semiconductor laser diode is probably the most versatile and useful of all the sources in fibre systems.

3.4 Optical sources – discussion

This chapter has considered in brief the principal sources used in fibre systems. For special applications there are many other possible light sources. These include dye lasers, which are particularly useful in that they produce high power (of the order of watts) and may be tuned over a range of wavelengths (usually in the blue to yellow region, and thus rather short for long-distance transmission in fibres). There are numerous other diode lasers, based on II—VI compounds rather than III—V compounds, which are useful in the few microns wavelength region. These may become important sources for future long-wavelength systems using halide crystal fibres; however, it should be mentioned that both source and detector require cooling, since the photon energy at a few microns wavelength is only a few times kT at room temperature.

As a 'broad brush' summary of source properties, both coherent and incoherent, Table 3.2 presents the more important source parameters. This table is intended as an outline guide only. There are numerous exceptions to all the very general statements therein, especially for semiconductor sources. In system design, it is therefore probably optimum to determine the system requirements, then look through samples of an appropriate looking source until a close approach is found. Examples of systems analysis to determine source properties are given in the chapters on sensing systems.

The preceding treatment of optical sources has concentrated on source properties and has, quite deliberately, omitted any detailed descriptions of device physics. The objective has been primarily to alert the reader to the important operational features of optical sources, and to provide a guide from which to locate the appropriate more detailed treatments in the literature. Interestingly, the complete books on sources – listed in references – all limit their treatment of these operational aspects of devices, so our description here, though brief, is complementary to these contributions.

Chapter 4

Light detectors

4.1 Introduction

The devices most commonly used as light detectors in fibre optic sensing systems are semiconductor diodes. Occasionally other devices, such as charge-coupled device arrays, photoconductors and photomultiplier tubes, may be required. In this chapter the basic principles of semiconductor light detectors and photomultiplier tubes (PMT) are very briefly reviewed, and the operational characteristics of these devices are analysed.

All these photodetectors rely on the energy of the incident photon either to produce an electron–hole pair in the semiconductors or to cause the release of a primary electron from the cathode of a PMT. Thus the detection process is in effect a photon counter, and as such has certain statistical properties determined by the fact that counting is involved. In particular, a phenomenon known as shot noise is inherent in all optical detectors of this type. Expressed at its simplest, if the incident optical power is equivalent to an average flow of N photons in some time interval τ, then if the number of incident photons in successive intervals τ is counted, that number will have a variance of $\sqrt{N}$. This corresponds to a fluctuation in the optical power observed in the interval τ of $\sqrt{N}$ photons, and thus one can define a signal to noise ratio in a bandwidth of $1/\tau$ of $\sqrt{N}$. The value of N is, of course, simply related to the optical power P through $P = Nh\nu/\tau$, where h is Planck's constant and ν is the optical frequency [4.1].

There are numerous reasons why only a fraction of the incident photons are converted to useful electrons in the detection device. This fraction is known as the quantum efficiency of the detector, and clearly the quantum efficiency is one of the most important parameters describing a particular detector. In a well designed device, figures of over 70% are common. These optical detection devices all rely on the photon energy exceeding some threshold energy – the bandgap in the case of semiconductors and the cathode work function in the case of a PMT. There is thus an optical wavelength above which the detector is nonresponsive, and this may be readily estimated using the requirement that:

$$h\nu > E_{\text{thresh}}$$

Alternatively, in terms of wavelengths we find that the requirement is:

$$\lambda \lesssim 1 \cdot 2 / E_{\text{thresh}}$$

where the wavelength λ is in microns and the energy E_{thresh} in electron volts. The quantum efficiency will typically peak at about three-quarters of this wavelength.

One final, and often neglected, aspect of photodetectors is that they are all square-law devices; that is, the input optical power is converted to an electrical current in the detection process. Thus the electrical detected power is proportional to the square of the optical power. There are situations in which this is critical, especially in the interpretation of optical and electrical signal to noise ratios.

There are numerous treatments of the basics of photodiodes [4.2] and of the specific characteristics of photodiodes for use in optical communications available in the literature [4.1–4.5]. There is one important distinction between the discussion relating to optical communications systems and that relating to sensors and kindred systems which are the subject of this book. That is that the received power levels are usually much higher, and the bandwidths typically much lower, than those used in communications. This makes the detection process simpler, but there are circumstances in which more complex detection devices may be required. The principal objective of this chapter is to define the performance levels available and to establish criteria for the use of various detection systems.

4.2 Semiconductor photodetectors

There are currently four principal types of semiconductor photodetector in common use in optical systems — the PIN diode, the avalanche photodiode (APD), the PIN-FET hybrid module and photoconductors. The basic detection process is identical in all these devices — the creation of electron–hole pairs by incident photons, and so the wavelength response of the detector is determined primarily by the bandgap of the semiconductor material. Table 4.1 shows the bandgaps of a number of commonly used semiconductors.

4.2.1 The PIN diode

The silicon PIN diode (P-type, intrinsic, N-type) is probably the most commonly used photodetector for other than long-haul communications applications. A cross-section of a typical device is shown in Fig. 4.1. The width of the intrinsic (I) region is designed to absorb the maximum practical amount of the incident light at the wavelength of interest. Most of the light absorbed in this region will produce electron–hole pairs which, when the junction is reverse biased such that the I region is completely depleted, will be swept to the contact regions by the electric field in the depletion layer. Any photons which are absorbed outside this region will also produce electron–hole pairs, but these do not make any significant contribution to the detected current, since they are not swept along by the electric field and therefore recombine.

Absorption in semiconductors at wavelengths around the band edge varies very rapidly with optical frequency. Fig. 4.2 shows the calculated absorption efficiency of silicon over the wavelength range 0·5 to 1·0 microns. A convenient width for the I region is from 10 to 20 microns, and from this we may infer the Si photodetectors will be at their most efficient in the 0·8 to 0·9 micron wavelength region − which coincides almost exactly with the wavelength range of GaAs laser diodes and

Table 4.1   *Properties of semiconductors used in photodiode detectors*

Material	Bandgap	Cut-off wavelength[1]	Ratio of ionisation coefficients[2]
	eV	λ_{max}, microns	α/β
Silicon	1·11	1·1	10−100
Germanium	0·67	1·8	0·5
Gallium arsenide	1·43	0·85	1−0·01
$Ga_{0·86}In_{0·53}As$[3]	1·15	1·1	0·25
$Ga_{0·47}In_{0·53}As$	0·75	1·6	5
InAs	0·33	3·8	
InGaAsP	1·34 to 0·78	0·92 to 1·6	

[1] The optimum wavelenth will be typically 0·7 to 0·9 of the cut-off wavelength
[2] The ionisation coefficient ratio is a function of electric field levels, doping concentrations and crystal orientation. The figures given here express the range of possibilities. For more details see, for instance References 4·5 and 4·15
[3] A wide range of bandgaps is available using quaternary III−V and related compounds. The examples given here are representative,

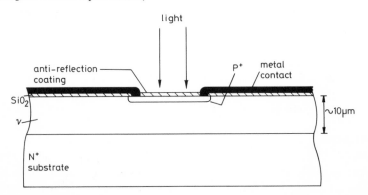

Fig. 4.1   *Section through a PIN diode*

LEDs. At shorter wavelengths, the efficiency falls because of absorption in the P contact region; if this is one micron thick it will absorb about 15% of the incident radiation at 0·7 microns, and at longer wavelengths the light passes through the I region with rapidly decreasing absorption. Clearly, the design of a given diode can be optimised for applications at unusual wavelengths, and very long I region PIN diodes have been used for detection at 1·06 microns wavelength [4.6]. However, the readily obtainable silicon PIN diodes are optimised for use at 800—900 nm wavelengths, with an I region of the order of 10 microns in width operating at reverse bias voltages of 10—50 volts. These devices are economic, have a very high performance and are simple to use. In systems requiring the use of longer operating wavelengths, either germanium diodes or III-V ternary or quaternary InGaAsP diodes may be used. The latter devices are currently optimised for use in high-speed long-haul communications networks [4.7], and general purpose diodes are not readily available. General purpose germanium diodes are easily obtained but are infrequently used.

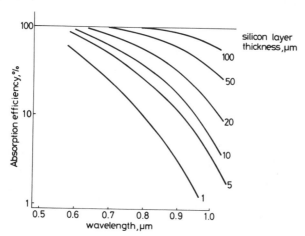

Fig. 4.2   *Absorption in silicon as a function of layer thickness and optical wavelength*

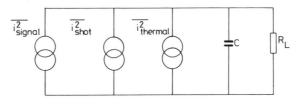

Fig. 4.3   *Equivalent circuit of a semiconductor diode photodetector*

The noise performance of a PIN diode may be understood with reference to the equivalent circuit model in Fig. 4.3. The current due to the incident optical signal may be written as:

$$i_{\text{sig}} = \frac{P_{\text{sig}}\eta q}{h\nu} \tag{4.1}$$

where η is the quantum efficiency and q is the electronic charge. The shot noise current due to the combination of the signal current and the dark current i_d is:

$$\langle i_{\text{shot}}{}^2 \rangle = 2q[i_d + i_{\text{opt}}]B \tag{4.2}$$

where B is the bandwidth and where i_{opt} is the *total* light-induced current, in contrast to i_{sig} which is the current induced by the optical signal sidebands carrying the information of interest. If the light is optimally modulated these two are the same, but, for instance, if the signal is a 1% amplitude modulation, the two contributions are most definitely different.

The final noise contribution is that due to thermal noise at the input to the preamplifier stages. This may be written as

$$\langle i_{\text{thermal}}{}^2 \rangle = \frac{4kTB}{R_L} F \tag{4.3}$$

where F is the noise factor of the input preamplifier, k is the Boltmann constant, T is temperature and R_L is the amplifier input resistance. The total electrical signal power to noise power ratio is then:

$$SNR = \frac{\langle i_{sig}{}^2 \rangle}{\langle i_{shot}{}^2 \rangle + 4kTBF/R_L} \tag{4.4}$$

The noise contribution may be divided into two components. Shot noise due to the incident light is a signal-dependent term, whereas thermal input amplifier noise and dark current shot noise are signal independent. The thermal noise may be reduced by increasing the amplifier input resistance R_L but at the penalty of maximum available bandwidth B, which is related to R_L through:

$$R_L = \frac{1}{2\pi BC} \tag{4.5}$$

where C is the detector capacitance.

It is a simple matter to estimate the ratio of thermal noise to dark current noise in terms of the bandwidth and the dark current density, to give:

$$\frac{\langle i_{\text{thermal}}{}^2 \rangle}{\langle i_{shot}{}^2 \rangle} \sim \frac{10^{-11} B}{J_d} \tag{4.6}$$

where J_d is the dark current density in A/cm^2. In silicon PIN diodes ($J_d \sim 10^{-9}$ A/cm^2), thermal noise will almost always dominate. In Ge ($J_d \sim 10^{-4}$ A/cm^2), dark current noise is correspondingly higher, though for systems of medium to large bandwidth the thermal noise will still dominate.

Thus, since the principal discussion here concerns silicon PIN diodes, we may neglect the effects of dark current noise in the *SNR* calculation, and thereby obtain the *SNR* as:

$$SNR = \frac{(P_{sig}\eta q/h\nu)^2}{(2qP_{opt}\eta q/h\nu + 8\pi kTBCF)B} \tag{4.7}$$

which makes the interesting, and important, point that the thermal noise performance of a PIN photodetector is limited by the detector capacitance, including the effects of any irreducible stray capacitances in the vicinity of the diode.

PIN photodetectors are usually operated in the so-called shot- (or photon-)

noise-limited regime, when the shot noise term in the above equation dominates. Substitution of values into the above gives as the condition for photon-noise-limited detection:

$$P_{opt} \gg 3CB \qquad \text{watts} \tag{4.8}$$

As an example, a simple inexpensive photodiode may have a stray capacitance of 10 pF so that, for a sensor application requiring a bandwidth of 100 kHz, a receive power of a few microwatts would ensure photon-noise-limited detection. Under these conditions, the *SNR* is approximately

$$SNR \sim (10^{19} P_{sig}^2 / P_{opt}) B \tag{4.9}$$

For the example discussed above this gives an *SNR* of approximately 90 dB, assuming that the power in the signal bandwidth is equal to the total optical power. It is also interesting to note that the minimum detectable power level, when the signal just equals the thermal noise level, requires P_{sig} to be approximately one picowatt, and P_{opt} to be sufficiently low for shot noise to remain negligible.

Signal to noise ratio calculations of this nature are vital in determining the design of the detector type for use in a particular application. For many sensor systems, light levels exceed a few microwatts, and bandwidths are below the megahertz region, so that a rule of thumb involving the use of a PIN diode with correct time constants built into the R_L values will give – usually – more than adequate performance. Some care is needed, even with these relatively simple applications, in the definition of the P_{sig} term to include *only* that power in the sidebands of interest around the optical carrier frequency. Other frequencies constitute nothing more than a source of shot noise. This discussion is continued in Chapter 5, where modulation schemes for use in optical sensor systems are described.

4.2.2 *The avalanche photodiode (APD)*

The silicon APD is an overall structure very similar to the PIN diode in Fig. 4.1. The same considerations concerning the optimisation of absorption in the depletion region also apply. However, the principal difference lies in the bias applied to the structure and in the details of the I region, which are optimised to encourage avalanche multiplication to occur [4.8]. The electric field levels in the depletion region are raised to the level that any electron–hole pairs created by incident photons will collide with the crystal lattice and create yet further electron–hole pairs as they traverse the depletion region. Thus the device has a gain associated with it, and the bias levels are chosen such that the gain is typically between 10 and 100. There are problems involved in exploiting this gain. The gain should remain stable over a range of temperatures; however, the ionisation coefficients, which determine the gain characteristics, are a strong function of temperature, so that the bias circuit for an APD requires some compensation for thermal drifts. The gain is also a very sensitive function of electric field (the ionisation coefficients are approximately exponentially dependent on the field) so that even for constant temperature the bias applied to the device needs to be kept constant to within very close limits (of the order of tens of millivolts). Finally, the absorption constraints impose the requirement of a long, low-doped depletion region, which consequently requires a large potential difference across it to raise the electric fields to the required level for multiplication. The total voltage is usually in the region of 200 V. The gain process is also statistical in nature. One created electron–hole pair may, on average, be multiplied by, say, 50. However, there will be a spread when the histories of individual electron hole pairs are traced [4.9]. This results in an excess noise factor introduced by the avalanche process.

By expanding the treatment for the PIN diode to account for the multiplication factor M in the current through the diode (which acts on both the optically induced and the dark current) and the excess noise factor, $f(M)$, we obtain for the perceived signal to noise ratio in the APD:

$$SNR = \frac{(P_{sig}q\eta/h\nu)^2 M^2}{\{[2qi_d + 2qP_{opt}q\eta/h\nu] M^2 f(M) + 4(kT/R_L)BF\}} \quad (4.10)$$

The behaviour of $f(M)$ is presented in various forms, and one convenient version is

$$f = M\left\{1 - (1-k)\left(\frac{M-1}{M}\right)^2\right\} \quad (4.11)$$

where the factor k is the ratio of ionisation rates of holes and electrons in the material [4.10]. The excess noise factor is thus reduced when k is small (as in the case for silicon) and is at a maximum when the ionisation rates are equal (which is almost the case for germanium and many III-V compounds – see Table 4.1).

In practice, the APD is used in situations where the shot noise is, before multiplication, well below the thermal noise. The optimum SNR is obtained when the multiplication process brings the shot noise up to the same level as the thermal noise. Further increase in M will cause the SNR to deteriorate. Thus the influence of the excess noise factor determines only the details of the multiplication level at which this optimum SNR is obtained.

APDs are principally used in long-haul telecommunications systems. Their applications in sensing and signal processing are limited to the relatively rare cases in which the received SNR is inadequate and is thermally limited. A simple analysis of the anticipated optical signal levels at the diode is necessary to decide whether the use of the quite complex photodiode and associated circuitry is essential to the performance requirements of the system.

4.2.3 The PIN-FET module

The PIN-FET module is a combination of a small-area, low-capacitance photodiode and a high-output-impedance FET preamplifier. The device is used as a compact hybrid assembly in which all lead lengths and stray capacitances are kept to the absolute minimum [4.11, 4.12]. Thus both the capacitance is low and the input impedance high, so that the effects of thermal noise may be kept at a minimum. The system has the advantages of low power supply voltages and stability over a range of temperature and electrical bias conditions. Comparable signal to noise ratios with the same values of P_{sig} are attainable with the PIN-FET and the APD, and the advantages become more pronounced at longer wavelengths where the silicon detector is no longer usable.

It is probably true that the PIN-FET, like the APD, will find most of its applications in the communications industry. However, in circumstances which may indicate that the APD is a suitable detector, probably the PIN-FET will be also. The detailed noise analysis of the PIN-FET essentially follows the reasoning underlying the PIN diode analysis, with the practical additional flexibility of a low diode capacity and an adjacent high-input-impedance amplifier.

4.2.4 *Photoconductors*

Photoconductive semiconductor detectors operate by an incident photon exciting an electron from the valence band into an acceptor impurity level, or, in N-type semiconductors, an electron from a donor level into the conduction band. After excitation, the resulting mobile carrier will increase the conductivity of the semiconductor sample for the duration of the carrier lifetime (or the transit time through the device if this is shorter). If the carrier lifetime is τ_0 and the incident optical power P_{opt}, then the average number of light-induced carriers N_{opt} will be:

$$N_{opt} = \frac{\eta P_{opt}}{h\nu} \tau_0 \tag{4.12}$$

The current through the photoconductor depends on the number of carriers and $\mathscr{E}$, the applied electric field. The field causes the carriers to drift at a velocity $v = \mu\mathscr{E}$, spending a time $\tau_d = d/v$ traversing the crystal (thickness d). It is a simple matter to show that the consequent photocurrent is

$$i_{opt} = \frac{\eta P}{h\nu} \frac{\tau_0}{\tau_d} q \tag{4.13}$$

An energy level diagram illustrating this process is shown in Fig. 4.4.

In contrast to photodiodes, the photoconductive detector requires a photon energy of only the difference between the band edge and the impurity level, which is usually a few tens of meV. Consequently photoconductors can be used as detectors well into the infra-red, out to perhaps 30 microns wavelength where the photon energy is 40 meV. It is, of course, necessary with any detector in which the photon energy is less than about 0·2 volts to supply cooling, since at room temperature the value of kT corresponds to 0·02 eV.

There are, of course, photoconductor detectors designed for use in the visible region, the most common being cadmium sulphide. These are inexpensive but now rarely used devices. Their relative instability and large size compared with silicon photodiodes has increased the usage of the latter device in applications, for instance photographic equipment, requiring modest performance at low cost.

Noise contributions in photoconductors stem primarily from the statistical nature of the recombination process — so-called recombination noise. The statistics of the recombination process bring variations in the current contribution per photon. A full analysis of noise in photoconductors is given in Reference 4.1.

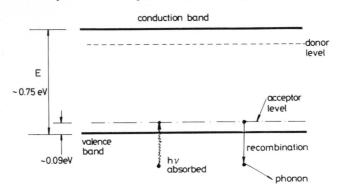

Fig. 4.4   *Energy level diagram representation of photoconduction. Here, an absorbed photon creates a hole in the valence band, which later recombines in a characteristic time — the recombination time — during which its energy is released as heat*

4.2.5 Charge-coupled device array detectors

Charge-coupled devices (CCDs) are becoming readily available for imaging applications, and it is natural that their application to specific problems in optical fibre sensors should be rapidly expanding.

Both line and area imaging devices may be obtained. The essential features of CCD arrays are that a charge proportional to the total optical power incident on a particular segment is accumulated and may then be read out in sequence under the control of a suitable clock system. A schematic of a linear array is shown in Fig. 4.5,

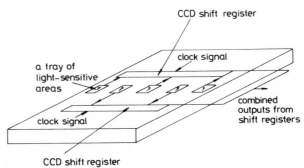

Fig. 4.5   *Schematic of CCD light-sensitive linear array*

and a full account of the device physics may be found in, for instance, Reference 4.13. Applications of CCD arrays include uses in the reading of spectrometers and in the analysis of spatial optical data. In both cases, the shift register read-out of the optical information transforms spatial information into a pulse train which is compatible with on-line digital processing. The CCD array is thus most attractive when multiple optical outputs from a sensor system are to be subjected to complex processing before the presentation of the finally resolved parameter to be measured.

4.3 Photomultiplier tubes

The photomultiplier tube is the most sensitive of optical detectors, and photon fluxes as low as one per second can be measured. The principle of the photomultiplier tube is shown in Fig. 4.6. An incident photon, provided that its energy

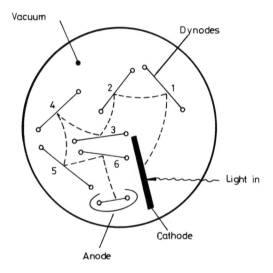

Fig. 4.6 *Diagrammatic representation of a photomultiplier tube*

exceeds the surface work function of the cathode, will cause the emission of a secondary electron from the cathode. The cathode material then determines the wavelength response of the detection process (see, for instance, Reference 4.14), and cathodes suitable for use from the near infra-red to the near ultraviolet are available.

The dynodes are biased at successively higher potentials, usually about 100 volts per stage, and the photon-generated electron is accelerated through the successive stages, and secondary emission from each stage multiplies the initial current. If each dynode generates N secondary electrons per incident electron, and there are M dynodes, then the overall gain of the PMT is N^M. Typically, N will be approximately 5 and M, 8 to 10, producing gains of from 4×10^5 to 10^7. The multiplied photon-induced electron current is finally collected by the anode.

The powerful feature of the photomultiplier tube is in its noise performance. In contrast to the APD, here the multiplication process is well controlled, and is

Table 4.2   Comparison of photodetector devices

Device type	Factors determining detection threshold	Typical operating power ranges[1]	Wavelength range microns μm	Quantum efficiencies	Frequency response
PIN[2]	Photon current exceeds thermal noise current	Usually shot noise limited. $P > 100$ nW typical	0·4 to 1·6, depends on material	Over 50%	Over 1 GHz available
PIN-FET[2]	Photon current exceeds thermal noise current	Operate thermal noise limited $P < 100$ nW typical	0·8 to 0·9, best at 1·3 and 1·5	Over 50%	Over 1 GHz
APD[4]	Multiplied photon current to exceed thermal noise current	Multiplication of 10–100 typical $P < 100$ nW	0·8 to 0·9, also available at 1·3 and 1·5	Over 50%	Over 1 GHz available
Photo-conductors[4]	Determined by recombination noise and thermal noise	Depends on detection scheme. High sensitivity possible (see Reference 14.16) ($\sim 10^{-18}$ W) in cooled device	To tens	Can be over 50%, often much less	100 MHz available
PMT[6]	Make multiplied photon current exceed thermal noise. Gain process almost noise free	10^{-19} watts can be detected. Always used at below 1 nW. High power can damage cathode	0·3 to 1·0	Low, typically below 10%	100 MHz possible

[1] These power ranges are purely for guidance – the individual circumstance should be carefully assessed for a particular application. In particular, the required bandwidths and signal to noise ratios within the required bandwidths can vary these figures considerably. The effects of $1/f$ noise can increase low-frequency noise by 20–40 dB
[2] The simplest photodetector in use. Low bias voltage, highly stable, inexpensive, readily obtained
[3] Low-capacitance PIN and high-impedance input FET give excellent sensitivity. Expensive component, manufactured by only a few companies. Simple to use and stable in operation
[4] High bias voltages requires, highly stabilised and temperature compensated. SNR reduces if multiplication set too high
[5] Photoconductors are very simple to use, except that the longer-wavelength devices require cooling
[6] Very high gains are possible, and dark currents are minute. A relatively convenient detector for high sensitivity and low noise

arranged such that the gain level is well below the maximum attainable from the tube. There is consequently negligible excess noise introduced. The noise levels are therefore determined by the shot noise of the multiplied signal:

$$\langle i^2_{\text{shot}} \rangle = 2q(i_{\text{cathode}} + i_{\text{dark}}) G^2 B \tag{4.14}$$

and thermal noise in the load resistance:

$$\langle i^2_{\text{thermal}} \rangle = 4kTB \, (F/R_L) \tag{4.15}$$

where i_{cathode} is the photon-induced current at the cathode, and G is the photomultiplier gain. F is the noise figure of the amplifier following the PMT.

4.4 Photodetection – discussion

This chapter has briefly reviewed the types of photodetector available for use in optical sensor and signal processing systems. The choice of photodetector in a particular application depends on the available optical signal power, the optical background level and the required signal to noise ratio. The latter factor is effectively determined by the required resolution in the optical signal. In principle the resolution and the noise level should be identical, but it is standard practice to allow of the order of 10 dB margin so that, for instance, a resolution of 1 part in 1000 would require a minimum *SNR* of 40 dB [4.14].

Table 4.2 summarises the properties and applications areas of the detection systems discussed in this chapter. There are inevitably some generalisations, but from this a number of useful guidelines can be extracted. In particular, for the majority of sensing applications a PIN photodiode is adequate. A second important general point is that in all detection systems there will be a pronounced increase in the noise level at low frequencies owing to so-called $1/f$ noise in the detector itself and/or the input amplifier stages. This begins to be important at frequencies less than 10–100 kHz, depending on the device, and can severely reduce the *SNR* to these lower frequencies.

The reliability and stability of photodectors is invariably good, provided that they are used within specified limits of optical power and environmental conditions. PIN photodiodes do exhibit small temperature coefficients, both in overall sensitivity and in the variation with wavelengths. In some circumstances this is important, but in most applications no corrections for these effects are necessary. In APDs and PMTs the gain is a function of bias voltage and, especially for the APD, temperature. These devices are usually applied to the more difficult detection problems, and so the additional complexity of temperature compensation of stable high-level bias voltages is acceptable. Photoconductors are invariably applied to detection problems in the mid to far infra-red, where diode detectors are not available. They thus require cooling for noise reduction, but again this constraint is acceptable when the particular properties of longer-wavelength radiation – for instance a high degree of coherence available from CO_2 lasers at 10·6 microns – are required by the system application [4.17].

Chapter 5

Demodulation of light

5.1 Introduction

The previous chapter described the basic principles of a number of photodetector devices. In this chapter the actual detection of modulated light is considered. The various techniques involved in the demodulation process are described and the factors which limit the resolution of each of these techniques are estimated.

In the applications which form the main theme of this book all the principal parameters which describe a light beam may be modulated, sometimes simultaneously. These include light intensity, optical phase, polarisation, frequency (via, for instance a Doppler shift) and spectral distribution (or colour). However, it should perhaps be emphasised that photodetectors are only capable of detecting optical intensity, so that all these properties of the light must be finally detected as a variation in the optical intensity. Thus, somewhere in the detection process, there has to be a conversion step to produce the modulation (unless it is already intensity) in a detectable form.

The procedure in the assessment of the various modulation techniques is identical. The process whereby the modulation is converted to a suitably coded intensity is described, then the *SNR* estimated, assuming that this conversion process is perfect. The principal imperfections in the conversion process are then analysed, and the impact of these on the achievable resolution is assessed. As a general point, it should also be mentioned that these estimates will assume that any modulation frequency is well away from the effects of $1/f$ noise. At low frequencies, this may reduce the potential resolution by up to 40 dB below that theoretically possible. However, it is also possible that the individual demodulation processes will introduce errors which exceed this level.

5.2 Intensity modulation

An optical fibre system involving intensity modulation of light during its passage along the fibre is shown schematically in Fig. 5.1. If the received signal power at the detector is 10 microwatts – a typical figure for a simple system – the *SNR* in a 10 kHz bandwidth with a 10 kΩ load resistance using a PIN diode detector at a wavelength of 850 nm is readily found to be in excess of 80 dB, even though the device is operated in a thermal-noise-limited regime. This implies a resolution in power level of better than 1 part in 10^7. This intensity resolution is more than adequate for the majority of both signal processing and sensor applications.

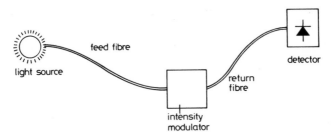

Fig. 5.1   *The general form of an intensity modulation fibre optic sensor*

However, the situation is complicated somewhat by the fact that intensity modulation may occur anywhere within the chain of components in Fig. 5.1. In particular, the mechanical coupling between source and fibre and detector may vary considerably, the components within the modulator itself may drift in relative positions with temperature, aging etc., the detector responsivity may vary with temperature and finally the source intensity may vary with aging and with noise on the electrical supply to the device.

Other more subtle effects may also influence the intensity of the light received at the output end. In particular modal noise, which occurs if the optical source is coherent and if the path from source to detector involves multiple paths [5.1, 5.2], can produce significant intensity modulation (a few per cent) owing to variations in the delay differences between the various modal paths – induced by, for instance, changes in fibre temperature or mechanical condition. Modal noise can itself be used as the basis for a range of possible sensors, and is also exploited in the 'fibredyne' data transmission system [5.3].

There are, consequently, two constraints imposed upon the design of a system involving intensity modulation techniques. First, the optical source must be incoherent; this implies the use of luminescent rather than any laser source, since multiple paths – often involved in single-mode systems as well as multimode – will produce an increase in base noise level even for path differences well in excess of the conventional 'coherence length'. Secondly, since intensity modulation is as likely to occur outside the sensor element as within it, some form of intensity reference is required. If the specified resolution falls outside the limits of a few per cent, this can probably be achieved by using a photodiode monitoring the launched light signal as a reference. The form of the reference will be discussed in detail in Chapter 6.

5.3 Detection of phase modulation

There are many optical systems which utilize the phase modulation of light as the means for transmitting information. The detection of optical phase modulation is usually performed via interferometric techniques. In an interferometer – for instance the Mach Zehnder shown in Fig. 5.2 – a sample of the light to be modulated is taken from the main beam by the beam splitter BS1 as a reference and later

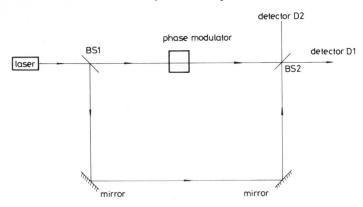

Fig. 5.2   *Mach Zehnder interferometer*

combined with the modulated beam by BS2. On the photodetector, D1, the fields from the two beams add, and the resultant is given by:

$$E_{\text{tot}} = E_1 \sin \omega_{\text{L}} t + E_2 \sin (\omega_{\text{L}} t + \phi(t)) \tag{5.1}$$

where E_1 is the optical field amplitude in the reference arm of the interferometer, E_2 is that in the modulated (signal) arm, and $\phi(t)$ is the time-varying difference between the optical phases in the two arms of the interferometer. The photodetector responds to the intensity of these combined beams, and thus the photodetector current $i(t)$ is

$$i(t) \propto [1 + \cos \phi(t)] \tag{5.2}$$

in diode D1 and proportional to $[1 - \cos \phi(t)]$ at D2. (For the case when the field amplitudes in the two arms are equal at the photodiode. See Reference 5.4.) The interference phenomenon then converts the *difference* in phase between the two beams to an intensity variation.

In radio frequency systems, the same function is usually performed using some form of coherent local oscillator. In optical systems, the required frequency stability for a coherent local oscillator which is not derived from the original optical source is of the order of parts in 10^{14}. Even though some progress has been made in this regard, and some success has been achieved with local oscillators in optical systems [5.5, 5.6] under most circumstances appropriate for fibre systems, it is impractical to detect phase modulation using anything other than interferometric techniques.

The response of the detected light intensity to small changes in relative phase between the two arms of the interferometer may be obtained by differentiating eqn. 5.2:

$$\delta i(t) = \sin\phi \, \delta\phi \qquad (5.3)$$

Thus the detected change depends on both the initial phase setting of the interferometer and on the change in phase. Clearly the change in phase is the important parameter, and this may be extracted conveniently if $\sin\phi = 1$ — that is, if the relative phases of the two arms of the interferometer are in quadrature. The relationship in the quadrature condition between the incremental phase change and the incremental intensity change is linear to within 1% over a range of 9°, to 2% up to 12° and to 10% up to 25° relative phase change. An interferometer with this quadrature bias between the two arms is thus, in principle, a useful technique for the measurement of optical phase.

There are, however, two principal difficulties with this arrangement. Maintaining a stable quadrature bias point is far from trivial, unless the interferometer itself is self-compensated in some way — as is, for instance, the Sagnac interferometer (see Chapter 7). Otherwise some form of automatic feedback technique [5.7], or a multiple source or similar arrangement, is required to read the state of the interferometer unambiguously. Both these are somewhat complex solutions to the problem. The other, perhaps more important, fundamental issue with this type of interferometer is that the output reading is sensitive to the optical power entering the interferometer as well as to the relative optical phase at the detector. A 2% drift in the source intersity produces exactly the same effect as somewhat more than $1°$ phase change. Source stability is thus critical, though it should be again emphasised that the stability is only critical in the bandwidth of the modulation applied to the interferometer. At other frequencies an intensity variation will manifest itself as a change in the interferometer scale factor (that is, the change in intensity per unit change in phase). Often these drift effects can be significantly reduced by adequate filtering.

There are other important distinctions to be made between interferometers designed for use with zero path differences (or an exact quarter wave to ensure correct quadrature bias), and those using larger path differences. In all finite path difference interferometers the effects of optical phase noise in the source are critical, since the path difference produces the correct conditions for conversion of the phase noise into amplitude noise. This is readily seen if we write the source signal as a $\cos(\omega t + \phi_n(t))$, where $\phi_n(t)$ represents a phase noise term. Then the two interfering beams, which will have a mean relative optical phase ϕ, will bear phase noise differences $\phi_n(t) - \phi_n(t - \tau)$, where τ is the time during which the light traverses the phase ϕ. The device is an interferometer, so that the output is now $1 + \cos(\phi_n(t) - \phi_n(t - \tau) + \phi)$. If the phase noise function can be filtered, then again this need cause little difficulty. However, laser sources are known to exhibit predominantly phase noise [5.9] especially at lower frequencies (below 0·1 to 1 MHz), so that conversion of phase noise to intensity noise may produce significant increases in background noise levels, and this increase will itself increase with increasing path difference. It is also frequently observed with laser sources that the phase noise spectrum contains more energy than the intensity noise spectrum out to much higher frequencies – in the gigahertz range at least. This noise power cannot be detected on direct intensity testing, but will manifest itself in an interferometer.

The other property of the source which is important in the use of interferometers with nonzero path differences is its wavelength stability – though perhaps one could argue that this is very low frequency phase noise? The importance of this may be gleaned from the consideration that the interferometer is really set to a time difference and not to an optical phase difference. In an interferometer in which the path difference is 10 mm, there will be rather more than 10^4 wavelengths of light from a semiconductor laser operating at 0·85 microns wavelength. If the operating wavelength changes by one part in 10^4, then there will be a change of one complete wavelength in the imbalance in the interferometer. This corresponds to changing the temperature of a semiconductor laser source by somewhat less than one degree! Wavelength stability is also important even in nominally zero path difference interferometers in that, without the use of elaborate detection techniques, a wavelength drift produces a corresponding fractional error in the perceived optical phase.

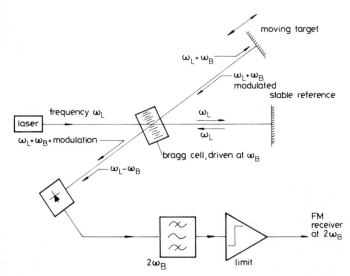

Fig. 5.3   *A heterodyning Michelson interferometer*

Many of these problems with interferometric detection may be overcome by the use of a heterodyne interferometer (see Fig. 5.3 for an example, a heterodyne Michelson interferometer). In the heterodyne interferometer, the reference beam is frequency shifted by passing it through a suitable modulator (in the diagram, the Bragg cell performs this function, imposing a frequency shift ω_B on deflected beams). It may be readily shown that for the case of equal amplitudes in the two arms, the output from the detector is $1 + \cos(2\omega_B t + \phi(t))$, where $\phi(t)$ is the difference in phase of the two arms. Thus, effectively, one-third of the total incident optical power is upshifted to the frequency $2\omega_B$. Usually this frequency is well outside the $1/f$ noise region, so there is often a gain in achievable signal to noise ratio of over 20 dB through this simple step. If $\phi(t)$ is periodic then this will modulate the intermediate frequency, and standard PM detection techniques may be used to extract the modulation. The spectrum at the intermediate frequency will consist of Bessel sidebands, and the minimum detectable level may be estimated by noting that the $J_1(\phi)$ sideband is approximately $\phi/2$ for small ϕ. Estimates based on the reception of a 1 mW local oscillator signal with a 100 kHz bandwidth indicate a theoretical signal to noise ratio of about 100 dB. Thus something in the order of 10^{-5} radians of AC phase modulation should be detectable. In practice, *SNR* of 80–90 dB is achievable, and minimum detectable levels in the region of 10^{-4} radians of phase modulation may be obtained.

The following summarises the interferometric detection process. Estimates of phase sensitivity of both quadrature homodyne and heterodyne interferometers indicate that resolutions of better than 10^{-8} radians should be attainable in bandwidths of the order of 1 kHz. In practice, these are achieved only with great care and skill in assembling the interferometer. However, microradians of sensitivity may be achieved fairly readily. For the homodyne interferometer, the effects of intensity noise on the source transform directly to spurios phase signals; for low-frequency work, $1/f$ noise in photodetectors can pose a significant drawback. For heterodyne interferometers, the intermediate signal is outside the $1/f$ noise bandwidth, and the signal is also electronically limited during the detection process, so that the effects of intensity noise on the source are to alter the *SNR* and not to alter the interferometer reading. For reasons concerned with phase to intensity noise conversion and with frequency stability of laser sources, short (preferably zero) path difference interferometers are preferred.

5.4 Detection of polarisation modulation

The characterisation of polarised light may be conveniently split into two areas — that in which the light is known to be in a particular class of states, for instance linear or circular, and that in which the state of polarisation of the light wave is totally unknown. The latter is a somewhat more complex problem, so initially we shall discuss the measurement of the orientation of linearly polarised light. At the outset is should be mentioned that, if the polarisation state of the light input to the polarimeter to be described is not purely linear, there will be errors in the reading. A description of the characteristics of the various states of polarisation is given in Appendix 5.

Fig. 5.4 shows a simple polarimeter suitable for determining the orientation of linearly polarised light. The Wollaston prism is a polarising beam splitter which separates the output directions of orthogonally polarised linear components of the input light. Thus the intensity I_1 in the figure is a measure of the linear component in a horizontal direction, and I_2 a measure of the linear component in the vertical direction. The orientation of the prism is arranged such that, when the input light is in its unmodulated position, the outputs from the two axes are equal. When the orientation deviates from this equilibrium position by an angle θ, then the amplitude of the components of the light on the 1 axis is $A \sin(\pi/4 + \theta)$ and on the 2 axis $A \cos(\pi/4 + \theta)$ (see vector diagram in Fig. 5.4). If it is remembered that the measured intensities are proportional to the square of these amplitudes, it is relatively straightforward to demonstrate that:

$$\sin 2\theta = \frac{I_1 - I_2}{I_1 + I_2} \tag{5.4}$$

This measurement is convenient in that it is purely ratiometric, and is independent of source intensity fluctuations and variations in the attenuation of the link between the source and the detector (unless, of course, the attenuation is polarisation

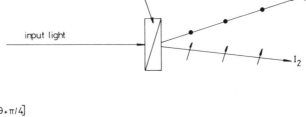

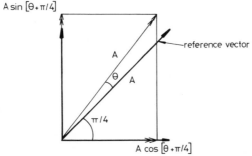

Fig. 5.4 *Use of a Wollaston prism as a linear polarisation analyser*

sensitive). For small angles the $2\theta = \sin 2\theta$ approximation may be used, and this function is linear to within 1% for angles less than 12°; at 20°, the deviation from linearity is 2%. Thus the actual orientation may be measured up to angles of 10° to 2% linearity without processing the resultant intensity signals. Further processing may be added to linearise the functions up to maximum angles of $\pm 45°$. For high-frequency measurements this processing may be inconveniently slow, though if the measurements remain stable over a number of cycles, averaging techniques are viable.

The intensity resolution is determined by the noise levels at the receive photodiodes. Polarisation-sensitive systems will almost always use highly polarised laser illumination, so that a receive optical power of at least $0.1\,mW$ per channel is probable. In a $100\,kHz$ bandwidth, the signal to noise ratio, with a $10\,k\Omega$ load resistor and a PIN detector, is found to be over $90\,dB$. Thus intensity resolution of one part in 10^8 is, in principle, feasible. This corresponds to an angular resolution in polarisation of the order of 10^{-6} degrees. This in turn implies a linear dynamic range of polarisation analysis of 10^7 to 1.

This, then, is the basic signal to noise ratio limitation of the receiver. Other factors conspire to reduce the resolution attainable by several orders of magnitude. In particular the rejection ratios of the polarisers – that is, the intensity of the light output from the horizontal output port when the input polarisation is purely vertical (or vice versa) – rarely exceeds $60\,dB$. This implies a resolution of the order of 10^{-2} degrees on the polarisation angle set by the separation ability of the polarisers. There are, inevitably, other optical components in the system to provide focusing of beams on to detectors etc., and these are virtually all, when measured at this resolution, birefringent. The errors accumulate, so that it is finally realistic to assume that polarisation resolution of 0.1 degrees may be readily obtained with a practical system, and 0.01 degrees with thorough characterisation of the optical elements in the measurement system coupled with on-line computation to remove these items from the measurement.

The problem of measuring more general states of polarisation is soluble using readily obtainable techniques [5.10, 5.11]. The techniques involve essentially the measurement of the so-called Stokes parameters of the incident polarised wave. To determine the Stokes parameters, the incident wave is passed through four devices which measure various polarisation properties of the wave. The devices are a polarisation-insensitive half-power attenuator, a linear polariser oriented horizontally, a linear polariser oriented at $45°$ to the horizontal, and a right-handed circular polariser. If the intensities of the light transmitted through these devices are I_0, I_1, I_2, I_3 respectively, then the Stokes parameters of the incident illumination are:

$$S_0 = 2I_0$$
$$S_1 = 2I_1 - 2I_0$$
$$S_2 = 2I_2 - 2I_0 \tag{5.5}$$
$$S_3 = 2I_3 - 2I_0$$

Examination of these parameters shows that S_1 is a measure of the amount of linear horizontal polarisation in the input beam, S_2 describes the amount of linear polarisation at $45°$ to the horizontal, and S_3 is a measure of the circularity of the incident beam. These parameters can be manipulated to completely characterise the state of polarisation of the input. As a set of simple examples, if we normalise such that $S_0 = 1$, we find that $S_1 = 1, S_2 = S_3 = 0$ corresponds to linear horizontally polarised light, while $S_1 = -1, S_2 = S_3 = 0$ is vertical linear polarised light. Similarly $S_3 = 1$, $S_1 = S_2 = 0$ is right circular polarised light and $S_3 = -1$, $S_1 = S_2 = 0$ is left circularly polarised light.

The Stokes parameters are a perfectly generalised description of the state of polarisation of any light wave, and impose no restrictions on the coherence of the radiation. There are other shorter descriptions of polarised waves (see Appendix 5) which require some degree of coherence before they may be rigorously applied. Particularly useful are the Jones vectors, which represent the wave as horizontal and vertical components, each described by a complex number denoting the relative phases between the two waves, and the Poincare sphere [5.13], which is a convenient geometrical representation of the progress of an arbitrarily polarised monochromatic wave through an arbitrary retardation element.

The general techniques for the determination of the state of polarisation of a light beam are limited in their resolution, not by available optical power, but again by uncertainties in the polarisation properties of the components intervening between the detectors and the input beam to be analysed. Careful measurements will, however, prove to be capable of similar resolution to those attainable for the linear polarisation analyser. The calculation of the actual state of polarisation is however now considerably more complex, and on-line computation facilities are necessary, in contrast to the linear polarisation analysis which could be performed entirely with analogue electronics.

5.5 Detection of optical frequency modulation

Optical frequency shifting is almost always associated with backscatter from moving targets introducing a Doppler shift. The frequency shift on the returned signal is readily calculated — for instance, with an HeNe gas laser operating at 633 nm wavelength, the Doppler shift is $1 \cdot 6 \times 10^6$ Hz/(m/s). Detection of the returned frequency-shifted radiation, and measurement of the frequency increment, are essentially interferometric, and there are two possibilities, heterodyne detection and homodyne detection. In the former case the returned signal is mixed with a sample from the radiated signal, and in the latter the reference signal is frequency shifted. The homodyne system has the great advantage of simplicity but cannot yield the sign of the Doppler shift, whereas the heterodyne system, at the expense of greatly increased complexity, gives complete information on the return signal.

There are many features of these detection systems which are common to phase measuring interferometers. In particular, the comments concerning the onset of phase noise with increasing path difference are again very relevant, and so is the important (often neglected) fact that the output is dependent on the frequency difference between the reference arm and the signal arm. Thus, again, all depends on the stability of the reference.

The criteria for frequency resolution in laser Doppler systems are discussed in some detail in Reference 5.13. The essential condition is that the offset returned frequency should exceed any anticipated drift in the frequency of the laser during the time difference between the signal and reference paths in the interferometer.

As a very rough estimate of the general trends, if we use the assumption that the anticipated root mean square (RMS) drift in frequency during the coherence length of the source is approximately the source linewidth $\Delta\nu_{1/2}$, and that the RMS frequency drift in a length l_1 less than the coherence length is a linear function, we arrive at the expression for the RMS frequency drift $\Delta\nu_l$:

$$\Delta\nu_l \sim \frac{(\Delta\nu_{1/2})^2}{c} l_1 \tag{5.6}$$

For an HeNe single-mode laser with a linewidth of 10 kHz, we obtain $\Delta\nu_l \sim l/3$ Hz. This estimate, though very approximate, does provide a useful guideline; for path differences well below one metre, frequency differences on one hertz are readily resolved, implying a velocity resolution of less than one micron per second.

The other criterion is that the returned signal be sufficiently strong to overcome the noise level present at that particular frequency in the bandwidth set by the receiver electronics. With relatively simple electronics, and laser powers in the milliwatt region, return losses of perhaps 70 dB can be detected. Increasing the frequency selectivity of the electronics, and using heterodyne techniques to avoid regions of $1/f$ noise, can increase the sensitivity to the region of 100 dB return loss. Thus, an optical laser Doppler probe can form the basis of a very sensitive instrument capable of excellent velocity resolution. Some applications are discussed in Chapter 8.

5.6 Detection of colour modulation

There are numerous optical measurement techniques which involve the measurement of colour, perhaps the best known being pH indicators. Measurement of colour electronically is, in its most general form, a difficult task, since it implies measurement of both the input light spectrum $f_{in}(\lambda)$ and the output light spectrum $f_{out}(\lambda)$ over the range of wavelengths of interest. The eye – and the colour TV camera – perform this function by observing light through appropriate red, blue and green filters. The use of an analogous technique in wavelength distribution measurements is possibly feasible, but presents the well known problem that different combinations of wavelengths may produce very similar subjective colours.

In most situations in which colour information is required, the problem is in fact considerably simplified. Often the output spectrum is all that is required, since the ratio $f_{out}(\lambda_1)/f_{out}(\lambda_2)$ provides the necessary information provided that the output effects at λ_1 and λ_2 are not dependent on the fine structure of $f_{in}(\lambda)$. The colour definition problem then reduces to the measurement of optical power at two different wavelengths.

There are two stages in the colour measurement process. The first is to perform the appropriate wavelength separation, and the second is to measure the optical power levels at these wavelengths. Wavelength separation may be implemented in a variety of ways, from complete prism or grating spectrometers [5.14] to interference filters [5.15] to simple coloured glass filters. Often the simplest solution is adequate, since the two wavelengths of interest may well be a considerable distance apart. There are also numerous wavelength multiplexing components becoming available for use in optical communications systems [5.16]. These are essentially spectrometers in a form which is compatible with simple interfacing with an optical fibre, and as such are especially convenient for use with fibre optic sensors. The wavelength resolution of these devices varies over a very large range depending on the device design, and this requires careful matching to the system requirements.

Received power monitoring is simply effected by use of photodetectors (usually PIN photodiodes) after the separation process. There are two practical factors to incorporate into the system design. The wavelength sensitivity of photodiodes varies between diodes, so that a characterised matched pair is usually required. For accurate relative intensity measurements, one of these diodes cannot be replaced without either replacing the other or completely recharacterising the pair. Secondly, the sensitivity variations with device temperature are wavelength dependent, so that careful thermal corrections are necessary to produce an accurate instrument. It is therefore virtually inevitable that any accurate colour-sensitive measuring instrument will have to incorporate some computer processing.

5.7 Discussion

It is interesting to note that in all the preceding detection schemes, more than adequate resolution in the parameter to be measured is available if optical and electronic noise were the only contributors to parameter uncertainties. However, in all cases there is a need for some form of reference level against which to measure the unknown. This is true for intensity, phase, frequency, polarisation and colour measurements. In the majority of systems it is the quality of this reference which limits available resolutions. Even so, there is always a more than adequate reliable dynamic range for the vast majority of sensor requirements. Perhaps signal processing imposes more stringent conditions, but this depends critically on the detailed requirements, and especially on whether the processing is to be analogue or digital.

Chapter 6

Intensity modulation transducers

6.1 Introduction

One type of optical fibre transducer utilises intensity modulation as the mechanism by which the information to be measured (the measurand) is transmitted on to the light guided within the fibre. There are therefore in effect two stages in the transduction process. The measurand is caused to interact with a light intensity modulator and this transfers the information on to the guided light. This involves either some form of mechanical motion of a mask or reflector interrupting the light path, or a so-called microbend transducer in which mechanical displacement is caused to produce a direct intensity modulation. There are isolated examples of other modulation mechanisms, some of which will also be described in this chapter.

The point that all forms of optical modulation must be – sooner or later – translated to an intensity variation for detection has already been made. However, it does sometimes lead to a rather hazy boundary between intensity modulated sensors and sensors utilising other modulation schemes. For instance, the black body temperature probe described in Chapter 9, and the polarisation modulated temperature measuring tip described in Chapter 10, could equally well be included as intensity modulated devices.

The format of this chapter, and also of Chapters 7–10, will be firstly to describe the modulation mechanisms available and to estimate the sensitivity of these mechanisms to environmental variations, and secondly to examine some of the possibilities by which the modulation phenomena may be interfaced with the desired measurand. There are also numerous practicalities to consider; in particular, the need for an intensity reference channel is paramount in sensors requiring stability of better than a few per cent. Intensity modulated sensors have the attraction of optical simplicity, but also have the drawback of mechanical complexity in the transformer from the measurand to the modulator. However, it could be strongly argued that the transducer industry excels in solving the mechanical problems, so that there will be considerable scope for the application of these skills to optical fibre devices.

6.2 Modulation mechanisms

Fig. 6.1 shows the general form of an intensity modulated sensor, and indicates the principal modulation mechanism — moving reflectors or transmitters and microbends. In the following analyses, we shall assume that the light source is a light-emitting diode (lasers are not convenient because of coherence effects especially modal noise) and that the received optical power at the detector is $10\,\mu\text{W}$. If the detection bandwidth is $10\,\text{kHz}$ and the detector load resistance is $10\,\text{k}\Omega$, then the *SNR*, for an input amplifier with a noise figure of 6, is 84 dB. The noise is thermally limited so that, if necessary, considerable improvement in *SNR* could result by using an APD (in fact another 40 dB is, in principle, attainable).

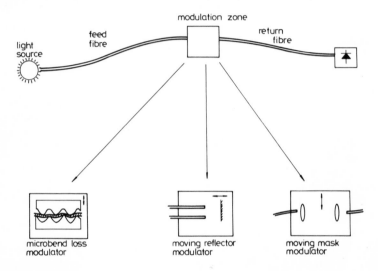

Fig. 6.1   *The basic forms of the three principal classes of intensity modulation sensors*

Therefore the fundamental limitation on achievable resolution — neglecting effects such as source noise, component drift, etc., all of which may be compensated — is in the order of 1 part in 10^7. In order to characterise the modulation mechanisms, we analyse the interaction between the modulation technique and the transmitted light intensity to determine the change in the modulator required to produce a detectable change in the transmitted intensity. The figures to be extracted are, as usual, representative of an enormous range of possibilities. In particular the bandwidth required by the measurand concerned is critical, and the resolution estimates must be interpreted accordingly.

6.2.1 External modulation – masks and reflectors
The simplicity of moving reflector modulators is shown in Fig. 6.2. The mirror is caused to move perpendicularly to the axis of the input and output fibres. The mirror forms a virtual image of the input fibre a distance d behind the mirror, so

that the response of the modulator is equivalent to a calculation of the coupling of a virtual fibre to the output fibre, separated by a vertical distance a from the input fibre. It is assumed initially that the fibres both have step index profiles with core diameters $2r$ and numerical apertures NA. We see immediately that no light is coupled into the output fibre for $d < a/2T$, and for $d > (a + 2r)/2T$ the output fibre intercepts a constant area πr^2 of the end of the cone of light emerging from the image of the input fibre (where $T = \tan (\sin^{-1} NA)$). The area of the bottom of the cone is $\pi (dT)^2$, so that the transmission of the gap over this range is $(r/2dT)^2$.

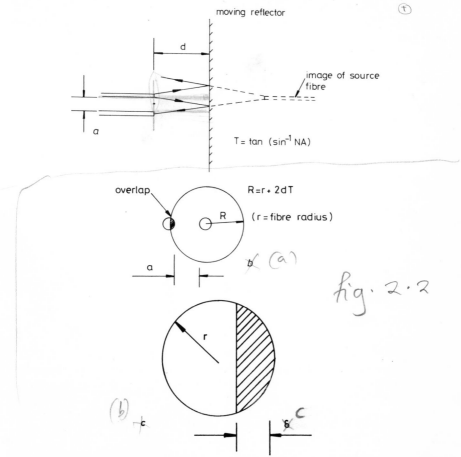

Fig. 6.2 *(a) Diagram of a moving reflector sensor coupling light between two fibres (b) The overlap between the light from the input fibre and the core of the output fibre determines the power coupled. (c) A simplified 'straight edge' model of the coupling process*

The amount of light coupled into the return fibre over intermediate distances is determined by the overlap of the cone from the image of the input fibre with the output fibre. This overlap, shown in Fig. 6.2b, may be calculated accurately using gamma functions [6.1] or one can use a linear approximation where the edge of the cone of light is approximated by a straight line across the output fibre.

In this case, simple geometrical analysis of the intercept area gives the fraction of the surface illuminated by the cone of light as:

$$\alpha = \frac{1}{\pi} \left\{ \cos^{-1} \left(1 - \frac{\delta}{r} \right) - \left(1 - \frac{\delta}{r} \right) \sin \left(\cos^{-1} \left(1 - \frac{\delta}{r} \right) \right) \right\} \tag{6.1}$$

where δ is the distance by which the cone edge overlaps the output fibre (see Fig. 6.2*c*). This is a most useful function, and is plotted in Fig. 6.3. These results

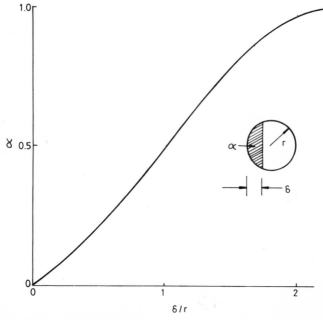

Fig. 6.3   *Numerical values for the 'straight edge' model showing fraction of the core area covered as a function of the position of the edge*

are not immediately applicable to the calculation of the variation of incident light intensity with reflector position, but this transformation is readily completed. The value of δ/r is calculated from the geometrical relationship:

$$\frac{\delta}{r} = \frac{2dT - a}{r} \tag{6.2}$$

and the fraction F of the incident optical power intercepted by the return fibre is simply:

$$F = \alpha \left(\frac{\delta}{r} \right) \left(\frac{r}{2dT} \right)^2 \tag{6.3}$$

These relationships provide complete design information for a reflection sensor of this type. As an example, Fig. 6.4 gives specific values for a pair of step index fibres, of core diameter 200 microns and numerical aperture 0·5, separated by a distance of 100 microns. The sensitivity of the device may be determined by taking the gradient of this function at its maximum, point A in the diagram — a separation of 200 microns. The rate of change of coupled energy with distance is here approximately 0·005% per micron. Assuming LED illumination and a total available power of 10 μW in 10 kHz bandwidth corresponding, as we have seen, to a resolution of 1 in 10^7, we find the intrinsic resolution of the sensor is better than 1 nm.

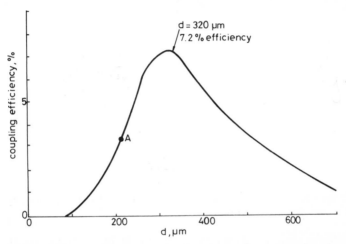

Fig. 6.4 *Calculation of coupling against reflector position for a sensor based upon a 200 micron diameter step index fibre with a numerica aperture of 0.5 spaced 100 microns from the illumination fibre*

There are a number of simplifications underlying this analysis. The fibre is assumed to have a step index profile, and also to be uniformly excited throughout its modal spectrum. This is implicit in the use of a uniform power density in the cone of light emerging from the fibre. A different index profile or nonuniform modal excitation will alter the result. (The power distribution in the far field from an optical fibre is used as the basis of a technique for the measurement of refractive index profiles [6.2] .) The assumption that the reflector is perpendicular to the fibre axes and has 100% reflectivity is also involved. Small tilting in the orientation of the reflector will have a small effect on the sensitivity, but will affect the separation corresponding to maximum sensitivity. Reflection losses produce corresponding reduction in the sensitivity.

The principles of a mask motion sensor are shown in Fig. 6.5*a*. The input fibre is imaged by the two-lens system on to the output fibre, and the mask travels perpendicularly to the direction of light propagation between the two lenses. The graph in Fig. 6.3 describes the percentage transmission of the device exactly (within limits imposed by the simplifications inherent in the previous analysis), and the sensitivity may be estimated as a change in δ/r of less than 1 part in 10^6. The simpler structure in Fig. 6.5b operates well without the lenses, but at a sensitivity reduced by the loss over the gap, which is $(r/dT)^2$; for instance, with a 200 micron $0.5\,NA$ core fibre and a 1 mm separation, this is a factor of 33, giving a resolution of about 0·1% of the core radius.

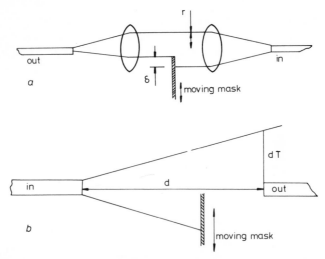

Fig. 6.5   *Illustrating the models used to calculate the coupling between two fibres used in simple mask displacement sensors*

The simple mask structure may be readily modified to increase the displacement sensitivity by, for instance, using two periodic masks, consisting of alternate transparent and opaque regions of equal width. The transmission through the pair of masks then varies from 50% when the masks completely overlap to zero when the opaque bars of one are completely over the transparent sections of the other. The output intensity is periodic in the total spatial period on the mask, and the resolution is in the region of parts in 10^6 of the spacing between the bars. This can form the basis of a very sensitive, and very simple, displacement sensor. The concept can also be extended, particularly in systems in which several feed and return fibres can be incorporated, to permit digital encoding of positional information, but with the limitation that the minimum bit interval is usually of the order of the fibre diameter.

Transducers operating on a similar intensity modulation principle may be implemented using a variety of other mechanical configurations. For instance, lateral or longitudinal motion of an input fibre with respect to an output fibre (see Fig. 6.6) is one possibility. The sensitivities of these devices may be calculated using an exactly similar procedure.

6.2.2 Internal modulation – bending loss

Microbending is a well known phenomena in optical fibre systems. It is caused by spatial variations in the lay of the optical fibre inducing coupling between the modes in the fibre. Some of the coupling is to radiative modes, so that micro-bending produces loss. A microbending loss modulator induces a controlled loss which may be accurately related to the position of the device causing the micro-bending.

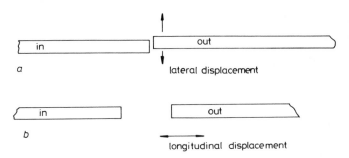

Fig. 6.6 *Simple sensor based on the position-dependent coupling between two fibres which move relative to each other (a) lateral (b) longitudinal*

The principle of the microbending loss modulator is shown in Fig. 6.7. A multi-mode fibre is sandwiched between a comb structure with spatial period Λ. This spatial period is chosen such that it matches the difference in propagation constants between suitably chosen modes in the fibre. The optical power distribution in the fibre is thus altered by this spatially induced coupling. In particular, some light originally propagating in the fibre core is transferred to the cladding (see Fig. 6.7*b*).

If the two modes to be caused to couple are designed by propagation constants β and β', then the required value of Λ is:

$$\Delta\beta = |\beta - \beta'| = 2\pi/\Lambda \tag{6.4}$$

For adjacent modes, the propagation constant difference is given by:

$$\Delta\beta = \beta_{m+1} - \beta_m = \left(\frac{\alpha}{\alpha+2}\right)^{1/2} \frac{2\sqrt{\Delta}}{a} \left|\frac{m}{M}\right|^{(2-\alpha)/(2+\alpha)} \tag{6.5}$$

where α is the grading constant of the fibre, and its radius, M is the total number of modes, m is the mode label and Δ is the refractive index difference between core and cladding. For parabolic index fibre, $\alpha = 2$ and

$$\Delta\beta = \frac{(2\Delta)^{1/2}}{a} \tag{6.6}$$

The required periodicity of the spatial modulator is then typically in the millimetre range.

The microbending transducer has been assembled both as a strain sensor [6.4] and as a dynamic acoustic sensor [6.5]. The sensitivity to the device may be charac-terised in terms of the rate of change of transmitted power with displacement, and, for an optimised device, a figure of the order of 5% per micron may be expected. Using our resolution criterion of 1 part in 10^7, we find that the displacement sensitivity is in the region of 0·01 mm. This sensitivity can be improved upon by increasing the optical power level at the receiver; for laser illumination of the fibre, a significant increase may be expected by reducing the receiver bandwidth and by increasing the interaction length of the sensor. Up to two orders of magnitude enhancement is thus possible, and this corresponds closely to values which have been achieved in practice.

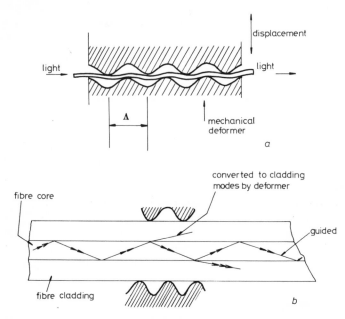

Fig. 6.7   *Principles of the microbending loss intensity sensor*

One other point concerning the microbend transducer is the fact that light displaced from the core may be guided in the cladding, and may be exploited in the detection system. Detection of cladding light is feasible using a simple spatial filtering arrangement, and if this is used as the signal the contrast of the light is increased at the detector. This can be an advantage if the detector is shot noise limited on the core light. It is also feasible to consider using the sum of the two channels – core and cladding light – as a reference for the signal level, which is the difference between the intensities in the two channels. This could compensate for intensity drifts.

6.2.3 Refractive index modulation sensors
A number of physical parameters cause changes in the refractive indices of materials. Temperature and pressure are the most important, though one could postulate

other effects. For instance, porous material could exhibit a refractive index which is strongly dependent on the chemical composition of a surrounding medium.

Perhaps the simplest system which uses variations in refractive index as the monitoring medium is step index fibre, in which the core glass and cladding glass have different refractive index temperature coefficients [6.6]. This forms the basis of an alarm system, since at the temperature at which the core and cladding indices become equal, the fibre ceases to act as a guide. The transition is clearly gradual, though the complete cut-off of the guide occurs at temperatures which can be defined to an accuracy of a few degrees. By using appropriately chosen glasses for the core and cladding, high- or low-temperature alarm systems can be designed. Applications include, for instance, low-temperature alarms in liquified natural gas storage tanks as well as conventional fire alarms. The safety aspects of the probe are particularly attractive.

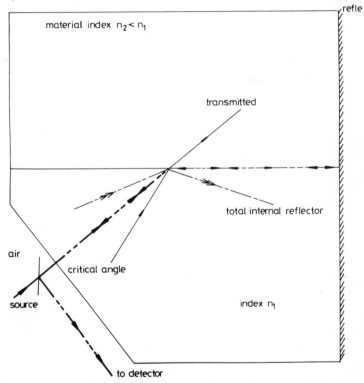

Fig. 6.8 *One form of total internal reflection sensor. Variations in n_2 produce variations in the power coupled back towards the source along the paths shown. The return is produced as a wide beam so that collection optics should also be incorporated*

Closely related to these sensors is a range of devices which involve the use of an optical fibre terminated by a material whose refractive index is sensitive to the required parameter, and arranged so that the incident angle of the light in the fibre at the interface is the critical angle.

The principle – in bulk optics – is shown in Fig. 6.8. At any dielectric interface, an incident light wave will produce both a refracted and reflected wave, and the angles of incidence and refraction are related through Snell's Law. Expressed in electromagnetic wave terms, Snell's Law involves matching wave vectors at this interface. When the incident wavefront is exactly at the critical angle, then the refracted wave travels along the interface. In the arrangement shown, the mirror will then reflect the refracted wave along the surface, and this will in turn be launched back – by reciprocity – along the incident path; if the incident angle is exactly the critical angle, then the detector will register a signal. The system then provides a very precise monitor of the relative dielectric constants of the two materials. Alternatively, one could arrange that the mirror is positioned to reflect the wave which is reflected from the dielectric interface.

The intensity reflection coefficients at the interface are given by the standard Frensnel reflection expressions [6.7] :

$$R_1 = \left| \frac{\cos\theta - (n^2 - \sin^2\theta)^{1/2}}{\cos\theta + (n^2 - \sin^2\theta)^{1/2}} \right|^2$$

$$R_{11} = \left| \frac{n^2 \cos\theta - (n^2 - \sin^2\theta)^{1/2}}{n^2 \cos\theta + (n^2 - \sin^2\theta)^{1/2}} \right|^2$$

(6.7)

for the perpendicular and parallel polarisation directions. The derivatives of these expressions give the rate of change of reflected power as a function of the refractive index n_2. An optical fibre sensor probe based on this principle has been described theoretically [6.8] but has, to date, only been reported experimentally in the bulk optical form.

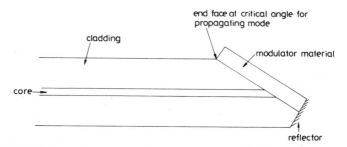

Fig. 6.9 *A proposed fibre optic total internal reflection modulation probe. Variations in the refractive index of the modulator material modulate the power coupled back into the single-mode feed guide*

The fibre optic total internal reflection probe is shown in Fig. 6.9. The difficulties with the practical implementation of such a probe are also apparent from the figure. The orientation of the reflecting surface with respect to the dielectric interface and the core of the fibre is critical. The fibre must be single mode, since multimode fibre has no defined 'ray' direction associated with the propagation of light within the core, and the effect is very strongly dependent on incident angles. Finally, the definition of critical angle for this interface will be modified significantly by the fact that the incident wave is in a guiding structure. The complete wave characterisation of the optical interactions at this interface is essential to ensure that the correct value of the critical angle is used. The boundary conditions at the dielectric interface must be recognised as those appropriate for guided waves in both the materials. The initial estimates of the sensitivity of this probe – in Reference 6.8 – indicate that it will be less sensitive to pressure changes than most other optical probes. However, it has great potential owing to its very small size.

6.2.4 Evanescent wave coupling sensors
Evanescent waves occur in optical fibres in two distinct situations. Directional couplers function by using evanescent waves as the coupling to transfer energy from one arm of the coupler to the other, and in situations where total internal reflection occurs – for instance at a glass–air interface – an evanescent field penetrates a short distance into the lower-index material.

Both these phenomena have been used as the basis of a sensing device. Fig. 6.10 shows the principles of a frustrated internal reflection displacement sensor.

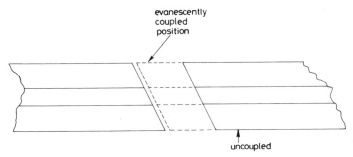

Fig. 6.10 *Conditions required for evanescent coupling between two fibres. The end faces must be cleaved parallel to each other, and must be spaced within a few optical wavelengths for significant coupling to occur*

The input and output fibres are coupled when the evanescent field from the input overlaps with that from the output fibre. The power coupled to the output is an extremely sensitive function of the relative positions of the two ends. The device has been configured as an acoustic sensor [6.9]. The alternative approach is to use coupling of the evanescent field between two waveguides, as in Fig. 6.11. Provided the two guides are identical, then varying the separation of the two varies the power coupled between them [6.10]. The device has been evaluated with fibres as the guides [6.11].

6.3 Intensity modulation techniques

Two general observations may be made concerning all the above modulation mechanisms. All (except the total internal reflection device) are sensitive primarily to displacement, and all require some form of reference intensity channel to ensure that drifts in components which occur with aging, thermal changes etc. are taken into account. Therefore there are two stages in the design of the actual transducer to exploit these techniques — the optimisation of a 'measurand to displacement transformer' and the incorporation of some form of reference channel.

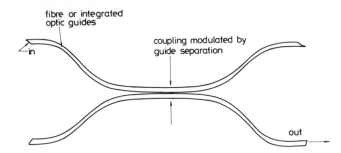

Fig. 6.11   *Displacement sensor based on modulation of the coupling between two waveguides. Coupling depends exponentially on the guide separations, so that displacement sensitivities of small fractions of a wavelength may be realised*

The transformer may take several forms. For acoustic transducers, appropriate acoustic design is clearly of fundamental importance in optimising the sensor sensitivity [6.12], though a simple reflecting diaphragm is often useful for static pressure measurements [6.13, 6.14] when interfaced with a reflection modulation system. Acceleration may be measured by using the displacement of a pendulum to actuate an optical fibre sensor [6.15]. Countless techniques have been evaluated, some of which are described in References 6.16 to 6.18, and the overall performances of some of the practical sensors are summarised in Chapter 11.

The majority of the techniques investigated experimentally have thus far ignored the problem of the reference channel. A notable exception is the so-called 'Fotonic' sensor, originally developed in the late 1960s. This is essentially a fibre bundle reflector system, in which half the fibres in the bundle are used to illuminate the reflector, and the other half to collect the returned signal. The basic characteristic of the device is then essentially as shown in Fig. 6.4, with some relatively minor detailed variations to account for the relative geometries of the transmit and receive optical fibre arrays [6.19]. There are a number of locations in which this sensor could suffer from intensity uncertainties, over and above those introduced by the motion of the reflector. The reflector itself could tarnish, the source could deteriorate with aging, the connector characteristics could change (though this is a minor problem with well designed optical fibre bundle systems), and some of the fibres within the bundles may break. With the exception of the last effect, these problems may be compensated for by the relatively simple expedient of using two sets of return fibres (see Fig. 6.12) displaced by a known amount from each other. Measuring the ratio of the returned powers from each of these two separate probes will give a unique value for the position of the reflector, provided that the correct portions of the characteristic in Fig. 6.4 are used.

The Fotonic device has been in use, in a few specialised applications, for some time. There are relatively few other reports of successful intensity drift compensation schemes in the literature. One suggested scheme is colour multiplexing,

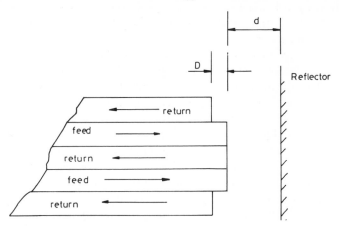

Fig. 6.12 *Reflection displacement sensor including some compensation for variations in reflector, optical coupling etc.*

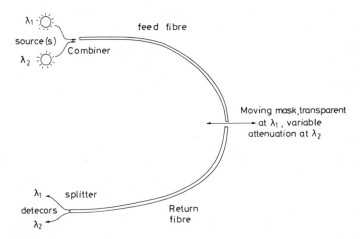

Fig. 6.13 *A wavelength multiplexed compensation scheme. The modulator varies the transmittance of one wavelength and not the other. The remainder of the optical system is assumed to affect both wavelengths equally*

described in Reference 6.20. The principle of this is shown diagrammatically in Fig. 6.13. The ratio of the intensities of the two wavelengths returned to monitoring point is a direct measure of the displacement of the sensor element (see also Chapter 9).

The effect of the intensity error needs to be carefully assessed for any given application. The usual difficulty is that the DC level of the light passing through the system is time dependent. If the parameter to be measured is itself very slowly varying, then this represents a drift in the reading; in particular, a false zero will often result. However, if the required signal is, for instance, at acoustic frequencies, then the intensity drift will produce a corresponding percentage error in the reading obtained and not a constant drift over all readings. Consequently, the drift error is usually more important if it occurs within the measurement bandwidth.

The principles involved in intensity sensors are well established. However, it is in the rapid expansion of available technology coupled with an appreciation of the wide range of modulation mechanisms available that significant progress may be made. As an example, the band edge effect in semiconductors has been postulated as a temperature measurement technique for some time. It is well known that the bandgap in silicon, for instance, decreases by approximately 2 mV per degree centigrade rise in temperature, and is little influenced by other parameters. The attenuation of light of photon energies close to the bandgap energy is a very strong function of wavelength, so that illumination of the semiconductor with light of the correct wavelength results in a temperature probe in which intensity transmitted is a function of temperature. However, the effects of intensity fluctuations are severe.

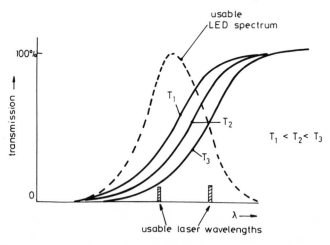

Fig. 6.14   *Band edge absorption thermometer, which may use LED illumination in its simplest form or two wavelengths of laser radiation in a compensated form*

If the attenuation against wavelength curve is plotted for a semiconductor sample, the shape of the curve varies relatively little, but its position changes with temperature (Fig. 6.14). Thus using two suitably chosen wavelengths around the band edge, coupled with wavelength multiplexing and demultiplexing optics — much of which has been developed for the telecommunications industry — enables a reliable ratiometric measurement to be made [6.21]. This is but one example of many implementations of the basic principles described in this chapter.

Chapter 7

Phase modulated optical fibre sensors

7.1 Introduction

Phase modulation of light as a highly sensitive monitor of environmental changes has been increasingly exploited over the past hundred years [7.1]. The principal attraction of optical phase modulation is its intrinsically high sensitivity to environmental modulation, so that very high resolution measurements are feasible. The incorporation of optical fibres into interferometers was initially suggested for the Sagnac rotation sensor [7.2] and, slightly later, for use in pressure, temperature and strain sensors [7.3].

The advantages of using fibres in interferometric sensors lie both in easing the alignment difficulties inherent in assembling interferometers with long arms, and in increasing the sensitivity of the phase modulation to the environmental parameter simply by increasing the optical path length exposed to the measurand. A suitably designed fibre interferometer takes optical interferometry from the laboratory optical bench into a compact and mechanically rugged piece of instrumentation.

Optical fibre interferometric sensors are among the most sensitive measurement devices yet evaluated. Hydrophones [7.4], magnetometers [7.5], accelerometers [7.6], strain gauges [7.7] and thermometers [7.8] have all been fabricated around the fibre interferometer, and all have achieved sensitivities exceeding that available from other technologies. Not only is the sensitivity high, but so is the achievable dynamic range. The intrinsic interaction — that is, the change of the measurand into phase variations — is, in all known cases, linear to the limits that it can be evaluated. This implies dynamic ranges for optical fibre interferometric sensors of the order of 150 dB. How much of this is actually available depends on the detection technique used to convert the phase modulation to a usable electrical signal.

The basic elements of an optical fibre interferometer are shown in Fig. 7.1. For reasons explained in Chapter 5, it is usually preferable to operate the interferometer with approximately equal length signal and reference arms to minimise the effects of laser phase noise. This also implies that the reference arm should be effectively shaded from the influences of the measurand, since both arms are equally sensitive.

The modulator is included in the reference arm to either incorporate the required 90° imbalance for homodyne detection or to provide the frequency offset for heterodyne detection. The beam splitters are most conveniently optical fibre 3 dB couplers, thereby eliminating mechanical noise common on discrete optical elements. The use of couplers does, however, limit the scope of the modulation function to phase modulators based on mechanical stretching of the fibre using piezoelectric elements, and this can prove to be inconvenient for some detection systems.

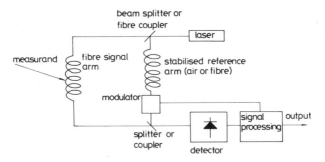

Fig. 7.1   *An optical fibre Mach Zehnder interferometer*

The remainder of this chapter examines the basic phase modulation mechanisms in both single-mode and multimode optical fibre waveguides, and continues to discuss two groups of optical fibre sensors, those involving the conversion of a mechanical effect into a phase modulation, and those using a purely optical effect which is not, to first order, influenced by the fibre properties.

7.2 Phase modulation mechanisms in optical fibres

The total phase of the light path along an optical fibre depends on three properties of the fibre guide:

> Its total physical length
> The refractive index and the index profile
> The geometrical transverse dimensions of the guide.

This assumes, of course, that the lightwave input to the fibre is monochromatic of wavelength λ_0 in air. (See also the Sagnac effect — Section 7.5.)

It is assumed that the index profile remains constant with environmental variations, so that all the following analysis concentrates on evaluating the depth of phase modulation for variations in length, refractive index and guide dimensions alone. These variations may then themselves be evaluated for a given perturbation applied to the fibre, and hence the phase sensitivity of the fibre to this perturbation can be estimated. The total physical length of an optical fibre may be modulated by:

    Application of a longitudinal strain
    Thermal expansion
    Application of a hydrostatic pressure causing expansion via Poisson's ratio.

The refractive index varies with:

    Temperature
    Pressure and longitudinal strain via the photoelastic effect

and the guide dimensions with:

    Radial strain in a pressure field
    Longitudinal strain through Poisson's ratio
    Thermal expansion.

Variations in environmental parameters other than pressure, strain and temperature must be converted to cause phase modulation. Thus the design of an optical fibre phase modulated transducers is in effect a two-stage process: the first stage involves the optics and the interferometer, and the second the mechanical interactions between the measurand and the modulation of phase in the fibre.

    The effect of a pure change in length, δL, on the phase of light propagating in the guide is readily determined:

$$\delta \phi_L = \frac{2\pi \delta L}{\lambda_g} = \frac{2\pi L}{\lambda_g} \epsilon_1 \tag{7.1}$$

where λ_g is the wavelength of the light in the guide and ϵ_1 is the longitudinal strain. The effect of a change in refractive index Δn is equally readily derived to be:

$$\delta \phi_n = \frac{2\pi}{\lambda_g} L \Delta n \tag{7.2}$$

in the limit that the change in λ_g is negligible. The change in guide wavelength with core diameter is a little more complicated to derive, but the general expression:

$$\Delta \phi_d = \frac{2\pi L}{\lambda_g^2} \frac{\partial \lambda_g}{\partial d} \Delta d \tag{7.3}$$

is clear from simple geometrical considerations. Derivation of $\partial \lambda_g / \partial d$ depends on the guide structure under consideration. The detailed mathematical formulation of this derivative for various types of guides is beyond the scope of this book, but the basis may be found in, for instance, Reference 7.9. It is, however, instructive to consider a simplified case to estimate the order of magnitude of the effect.

    Fig. 7.2 shows the transverse section of a dielectric guide of width d supporting a mode of order m. The relationship between the mode number and the guide width is simply:

$$d \cos \theta = m\lambda/2 \qquad (7.4)$$

The guide wavelength is:

$$\lambda_g = \lambda \sin \theta \qquad (7.5)$$

where λ is the propagation wavelength in the medium of the dielectric guide, refractive index n; that is,

$$\lambda = \lambda_0/n \qquad (7.6)$$

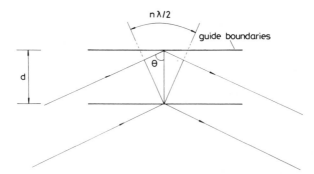

Fig. 7.2   *Transverse section of a dielectric guide. The boundary conditions are simplified to incorporate perfectly reflecting interfaces, but this alters the details of the analysis, but not the overall conclusions*

Hence we may arrive at the required derivative as:

$$\frac{\partial \lambda_g}{\partial d} = \frac{\partial \lambda_g}{\partial \theta} \frac{\partial \theta}{\partial d} = \lambda \frac{m\lambda}{2d} \frac{m\lambda}{2d^2 [1 - (m\lambda/2d)^2]^{1/2}}$$

$$= \frac{m^2}{4} \left(\frac{\lambda}{d}\right)^3 \frac{1}{[1 - (m/2d)^2]^{1/2}} \qquad (7.7)$$

giving, as an estimate for the phase modulation due to changes in diameter [7.10],

$$\Delta\phi_d = \frac{2\pi L}{\lambda} \left(\frac{m\lambda}{2d}\right)^2 \frac{1}{[1 - (m\lambda/2d)^2]^{1/2}} \frac{\Delta d}{d} \qquad (7.8)$$

The remainder of the interaction between an applied parameter and the optical phase change may then be characterised by deriving a suitable mechanical model to determine dimensional and refractive index changes. The simplest is the thermal variation in phase. Temperature changes the refractive index and the geometrical dimensions of the fibre. The change in phase is thus

$$\Delta\phi = \frac{2\pi L}{\lambda_g} \left\{ \alpha + \frac{\partial n}{\partial T} \right\} \Delta T \qquad (7.9)$$

since the change in diameter produces a negligible change in the propagation constant of the guide. For a pure silica guide, the temperature coefficient of expansion α is $5\cdot5 \times 10^{-7}/°C$ and the temperature coefficient of refractive index is $0\cdot68 \times 10^{-5}/°C$, resulting in a total phase variation with temperature of 106 radians per metre of interaction length. The actual figure observed in practice will vary somewhat with the detailed constituents of the optical fibre, including the effects of both doping and the core and cladding glasses and of any coatings on the fibre itself, which will modify the average coefficient of expansion of the composite.

The effects of a pressure variation on the optical phase path may be written in the most general form as:

$$\Delta\phi = \frac{2\pi L}{\lambda_g}\left\{\epsilon_1 - \frac{n^2}{2}(P_{11} + P_{12})\epsilon_r + P_{12}\epsilon_1\right\} \tag{7.10}$$

where ϵ_1 is the longitudinal strain and ϵ_r the radial strain. Again, it transpires that effects due to variations in diameter on the propagation constants in the guide can be neglected (the relevant numbers will be inserted later). P_{11} and P_{12} are the photoelastic constants of the silica fibre. A detailed derivation of the two strain values depends on the fibre configuration. It is useful to consider two extreme cases, corresponding to a longitudinal strain of zero (axially constrained) and a free fibre case (axially unconstrained)[7.11].

The fundamental static stress–strain relationships are:

$$\epsilon_x = \frac{1}{E}(S_x - \nu(S_y + S_z))$$

$$\epsilon_y = \frac{1}{E}(S_y - \nu(S_x + S_z)) \tag{7.11}$$

$$\epsilon_z = \frac{1}{E}(S_z - \nu(S_x + S_y))$$

where E is Young's modulus, ν is Poisson's ratio, S are the applied stresses and the subscripts x, y, z refer to coordinate axes. The fibre strain distribution has cylindrical symmetry, in which case $\epsilon_y = \epsilon_x = \epsilon_r$ and $\epsilon_z = \epsilon_l$. The static relationships apply when the applied pressure wave has a wavelength in the silica fibre which is much greater than the dimensions of the interaction region. This may be referred to as the axially unconstrained case. Manipulating the above expressions then gives for the relationship between ϵ_r and ϵ_z:

$$\epsilon_z = \frac{-2\nu}{1-\nu}\epsilon_r \tag{7.12}$$

for the application or a radial stress. The radial strain may be related to the pressure wave through the fundamental relationships above.

For cases in which the static approximation may not be used the situation becomes much more complex, and a full geometrical analysis of the acoustic interaction with the fibre is required. This involves specifying the acoustic boundary conditions, defining the input wave and computing the resultant strain distributions, from which the phase modulation depths may be obtained. Usually this will require numerical solution, but some cases are amenable to algebraic analysis.

The algebraically tractable situations include certain circular structures which may be approached using standard acoustic techniques [7.12]. Similarly rectangular structures may also be analysed, but these are rarely met in optical fibre sensors except perhaps as diaphragms. The simplest limiting case of this is that of the axially constrained fibre, in which the acoustic wavelength is much less than the interaction length and the applied pressure wavefront is plane and parallel to the fibre axis (see Fig. 7.3). In this case $\epsilon_z = 0$.

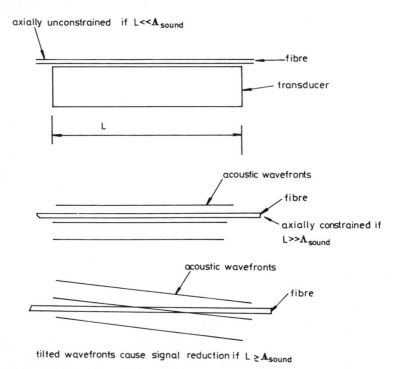

Fig. 7.3   *Geometry demonstrating the axially constrained and axially unconstrained limits of the acousto-optic interaction in fibres*

For the application of a pure axial strain, the radial strain is readily found from the Poisson coefficient of the fibre to be:

$$\epsilon_r = -\epsilon_l \nu \qquad (7.13)$$

where ν is Poisson's ratio.

There are clearly many other possible strain configurations, especially for situations in which the wavelength of the pressure wave is comparable with the radial dimensions of the fibre or with the dimensions of the form into which the fibre is configured. The former induces as asimuth-dependent strain configuration [7.13], consequently producing birefringence, and the latter is the source of numerous structural resonances.

The exact numerical values for the various mechanical constants are to some extent dependent upon the type of fibre considered. For pure silica, the mechanical constants are:

Young's modulus $1 \cdot 9 \times 10^{11}$ N/m^2
Poisson's ratio $0 \cdot 17$
Photoelastic constants P_{11} $0 \cdot 126$
$\qquad\qquad\qquad\qquad P_{21}$ $0 \cdot 274$
Refractive index $n = 1 \cdot 458$.

Inserting numerical values into the above expressions, we obtain for the phase sensitivity of an optical fibre transducer:

106   radians per degree C temperature change
10   radians per bar pressure change
11·4 radians per longitudinal microstain.

The function describing the effect of changes in diameter on the propagation constant is plotted in Fig. 7.4 for the case of a fibre of 50 microns guide width and a guide wavelength of 0·6 microns (corresponding to a free-space wavelength of 0·9 microns) for mode numbers of up to 100. The maximum mode numbers as a function of guide numerical aperture are also shown on the figure. These effects only occur in situations in which radial strain is present, and in these situations there are photoelastic effects which occur simultaneously.

The ratio of the guide perturbation factor $(m\lambda/2d)^2 \{1 - (m\lambda/2d)^2\}^{1/2}$ and the photoelastic factor $(n^2/2)(P_{11} + P_{21})$ determines the relative importance of these two effects. The latter is approximately 0·34 for silica, so that the guide perturbation factor may be neglected for guide numbers less than 30, which include all guides up to an *NA* of 0·2; even for a 0·5 *NA* guide, the guide correction factor on the highest-order mode is about 30% of the photoelastic effect.

The relative figures should be interpreted with some care, since the appropriate wavelengths must be inserted into the expressions to derive the correct total phase deviation. In particular the guide wavelength is a function of mode number; thus for high-*NA* step index guides there will be a considerable difference between the phase modulations on low-order and high-order modes, owing to the combination of the increasing influence of the guide perturbation with wave number. In graded index guides the situation is much less specific. The guide wavelength is approximately constant with mode number, and the equivalent expression for the rate of change of wavelength with radial strain, ϵ_r, in a parabolic index medium is derived from Reference 7.9:

$$\frac{\delta\lambda_g}{\lambda_g} = \left\{1 - \left(\frac{m}{M}\right)^{1/2} 2\Delta\right\}^{-1} \left\{\left(\frac{m}{M}\right)^{1/2} \Delta\right\} \epsilon_r \qquad (7.14)$$

Since Δ is typically 0·01, the effects of diameter variations may be ignored in most cases for multimode graded index guide.

These differences in modulation depth as a function of mode number are occasionally important, especially in systems which utilise mode–mode interference effects such as the Fibredyne data collection highway [7.14] and related sensors [7.15, 7.16]. However, in general, the effects of differential phase modulation may be neglected.

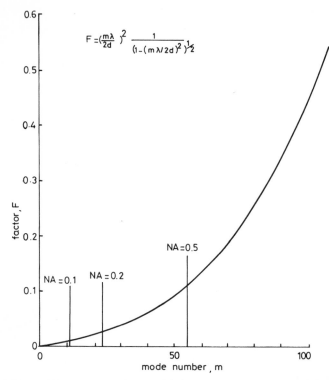

Fig. 7.4   *The correction factor F which represents the differential phase modulation between extreme modes in a multimode step index fibre. The mode number M is a measure of the numerical aperture.*

The inherent sensitivity of the silica fibre to variations in temperature, pressure and strain may be calculated by assuming a one microradian detection threshold etc. as outlined in Chapter 5; this gives, for a one metre length of fibre, detection thresholds of:

10^{-8} degrees
10^{-7} microstrain
10^{-7} bar (approximately 10^4 micropascals).

These figures indicate that temperature may be the most important variable. This is especially true since variations in temperature in homodyne interferometers will cause a drift in the bias point. For instance, a 10 metre length of fibre will suffer a phase change of $\pi/2$ for a temperature change of just over 1 millidegree. This may be compensated by feedback techniques or it may be removed using heterodyne detection, in which case the criterion is that the temperature variations may be filtered out. This is usually possible since the rate of temperature changes is significantly lower than the rate of signal changes. Other techniques — balanced interferometers — are needed if this requirement is not met.

Finally, it is interesting to evaluate the sensitivity of the fibre to acoustic power incident on a one metre length for the application of a pure strain and a pure hydrostatic pressure at the same acoustic frequency. The acoustic power density P_{ac} may be estimated from:

$$P_{ac} = \frac{p^2}{2Z_{ac}} \tag{7.15}$$

for an alternating pressure p, and for an alternating strain ϵ

$$P_{ac} = \frac{\omega^2 \epsilon^2 Z_{ac}}{2} \tag{7.16}$$

The total power requirement is the power density times the appropriate interaction area, the end faces for the strain cases and the total longitudinal area for the pressure case. The ratio of 'pressure power' to 'strain power' is then:

$$\frac{P_{ac\ pressure}}{P_{ac\ strain}} = \frac{p^2 (\pi r l)}{\omega^2 \epsilon^2 Z_{ac}^2 (\pi r^2)} \tag{7.17}$$

when calculated at the detection threshold. This estimate does, of course, implicitly assume that the interaction length with the fibre is significantly less than an acoustic wavelength, and some modification would be required if this were not the case. However, the ratio is so large that it emphasises a very basic point with all types of fibre phase sensors designed to monitor mechanical perturbations. This is simply that, if at all possible, it is preferable to ensure that the mechanical perturbation appears as a longitudinal strain rather than as a hydrostatic pressure variation, and this is probably true up to frequencies in the megahertz region. This has a fundamental bearing on the design of acoustic sensors [7.17].

An alternative approach is to examine the total amount of energy required to impose a given phase change — say ten radians — on a given length of fibre — say one metre — when the source of the energy is a radial force, a longitudinal force and a thermal change. If A_1 is the area of the surface of a one metre length of fibre, and A_2 is the area of the end face, then the ratio of work done to yield the same phase change in the form of an applied longitudinal force to an applied radial force is readily shown to be the order of $2\nu A_2/A_1$. This ratio is typically 10^{-4} to 10^{-5}. This emphasises that, at least for slowly varying perturbations, it is much more efficient to apply the perturbation as a longitudinal strain than as a radial strain.

A similar comparison may be made for the differential between the required thermal energy per unit phase change and the required longitudinal strain energy per unit phase change. In this case, it is readily demonstrated that the energy requirements for thermal input and longitudinal strain inputs are approximately equal.

This discussion on phase modulation neglects the Sagnac effect which is covered in detail in Section 7.5.

7.3 Optical fibre interferometers

An optical fibre phase-sensitive device requires some form of interferometer to perform the phase detection process. The exact format of this depends on the parameter to be sensed, and here the emphasis will be on environmental detection of pressure, strain and temperature. Rotation sensing is considered independently in Section 7.5.

A laboratory bench form of an optical fibre Mach Zehnder interferometer is readily implemented and shown in Fig. 7.5. This configuration is adequate for laboratory test purposes, but has a number of practical disadvantages when considered for use in other environments. They are:

(1) The physical size of the optical bench. This may be reduced by compact custom machining, but is still significant.
(2) All the components in the system external to the fibre are mechanically sensitive, and can introduce vibration noise. This can seriously deteriorate perceived signal to noise ratios.

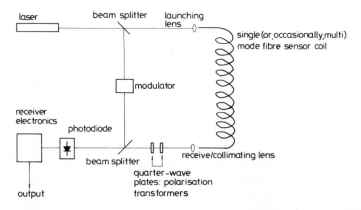

Fig. 7.5   *A fibre optic Mach Zehnder interferometer designed for use with an air path reference beam*

The laboratory model also has advantages; in particular, the format is easily changed and readily available optical components can be incorporated. This is particularly true for heterodyne interferometers, where a Bragg cell is by far the most convenient form of an optical frequency shifter. It is also fairly simple to modify parameters such as the path imbalance between the two arms to investigate source coherence effects.

For a practical sensor, an all-fibre interferometer is preferable – provided that suitably stable 3 dB couplers are available. The format of an all-fibre Mach Zehnder interferometer is shown in Fig. 7.6. There are numerous variations on this basic theme [7.18]. The advantages of the all-fibre format are in mechanical stability, depending critically on the coupler technology. A number of coupler techniques have been evaluated, for instance the 'bottle coupler' [7.19], the 'etched fused coupler' [7.20] and the 'block mounted couplers' [7.21] (see Fig. 7.7). The last of these is the most convenient laboratory format, but the component is bulky. In the longer term, integrated optics with a suitable interface to the fibre may be required to produce a rugged unit.

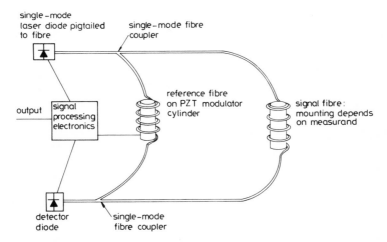

Fig. 7.6 *An all-fibre Mach Zehnder interferometer using homodyne detection incorporating the required quadrature bias via modulation of the feedback voltage to the PZT modulator*

The limitation on the all-fibre format is in the inconvenience in ensuring the stability of the operating point of the interferometer – that is, in maintaining the quadrature condition between the two arms. This has been successfully implemented by using a PZT cylinder energised via a feedback loop, but this system has the inherent disadvantage that the length modulation range of the PZT is, at most 2π; there is thus a resetting transient when the phase differential to be corrected drifts outside this value. Temperature drifts are the prime cause of relative phase fluctuations, and only very small temperature differences can be handled without the onset of noticeable transients.

A heterodyne system would be preferable. However, it is fundamentally impossible to incorporate the conventional Bragg frequency shifters into the all-fibre form without breaking the fibre and introducing further mechanically sensitive interfaces. A number of schemes have been suggested for turning optical phase modulation inteferometrically into a frequency shift [7.22, 7.23, 7.24], but none has been evaluated for an all-fibre device. With the exception of the coupled mode device [7.24] all these frequency shifters require the use of quadrature biased interferometers, so that the initial problem remains. The coupled mode device requires the use of a travelling controlled coupling point between two waveguides of different guide wavelengths. It is not at present clear how — or if — this could be incorporated into an all-fibre form. All these modulators are compatible with integrated optics, and again, the importance of a convenient and reliable interface between fibre and integrated optic guides becomes apparent.

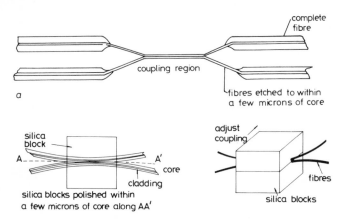

Fig. 7.7   *Optical fibre couplers for monomode systems (a) an etched coupler, and (b) the block coupler*

It is straightforward to impose a sinusoidal phase modulation on to light in an optical fibre, simply by using a PZT modulator, usually in cylinder form. Some degree of signal processing is possible with this. For instance the double-delay interferometer of Reference 7.18 effectively uses a quarter-wave delay at the frequency of the alternating disturbance along with a PZT sinusoidal modulator to modulate the disturbance on to the frequency applied through the PZT modulator. Another varient is the so-called 'synthetic heterodyne' [7.25] technique which uses the fact that if a sinusoidal phase modulation ϕ_m is applied to the reference arm, the signal arm will contain the signal

$$\nu_1 = J_1(\phi_m) \sin 2\omega_{sig} t \sin \phi \tag{7.18}$$

at the first harmonic of the signal frequency, and

$$\nu_2 = J_2(\phi_m) \sin 2\omega_{sig} t \cos \phi \tag{7.19}$$

at the second harmonic, where ϕ is the relative phase of the two arms [7.26]. Thus one or other of these two signals is always available.

Successful decoding of the signal is difficult, apart from the case of digital modulation, but it is in principle feasible. There are a number of other possibilities, some covered in outline in Reference 7.27. However, it is probably accurate to state that a truly satisfactory solution to the detection problem in an all-fibre interferometer remains to be discovered.

The preceding discussions have applied to the case of a single-mode all-fibre interferometer. In all cases, there are two orthogonally polarised eigenmodes, with different propagation velocities. The velocity difference is itself a function of temperature strain etc., so that the fibre acts as a variable-retardation plate. This implies that the state of polarisation of the output has a tendency to drift with time. This effect may be used as the basis of a sensor element [7.28] but is more usually experienced as an occasional fade in the interferometer output. The use of a variable-retardation plate or, for the all-fibre case, a polarisation controller [7.29] to optimise the initial response of the system is often adequate to ensure that a reasonably stable signal is maintained. This is particularly true for normal single-mode fibres in which the propagation velocities of the two modes are closely matched, so that the difference is a very slowly varying parameter. For highly birefringent fibres there will, in general, be a more pronounced drift effect unless the input light is oriented with the principal axes of the fibre.

In multimode fibres, several additional effects may also occur. The depth of the phase modulation on each mode in the fibre is somewhat different, to an extent depending on the doping profile of the fibre. This differential phase modulation may itself give rise to localised self-interference effects in the output pattern — the so-called 'Fibredyne' process, sometimes also known as modal noise. The modes may also be mixed with a reference beam [7.16, 7.30, 7.31] to give a resultant signal similar to that obtained with a monomode system, but with an accompanying fluctuation in the demodulated output caused by the random relative phases of the interfering modes and the modulation depth variations. Most of the spatial patterns of the output modes are spatially orthogonal to the reference beam, so that a large heterodyne mixing loss is inevitable. Even so, acceptable performance is attainable with a multimode system using demodulation via a suitable heterodyne reference beam.

The Fibredyne process has also been utilised as the basis for a number of sensor elements. The essential features of this interference process may be gleaned from a simple analysis of the interference of two waves, each of the same amplitude but slightly differently phase modulated and with relative phase ϕ_{12}. A square-law detector then receives:

$$P = A^2 [\sin (\omega t + \phi_{01} \sin \omega_m t) + \sin (\omega t + \phi_{02} \sin \omega_m t + \phi_{12})] \quad (7.20)$$

This reduces to:

$$P = (2J_1 (\phi_1 - \phi_2) \sin \phi_{12}) \sin \omega_m t \quad (7.21)$$

at the modulation frequency. The output is then non-linear for large relative phase deviations, and also tends to fade depending on the value of ϕ_{12}. However, for sufficiently large relative phase deviations at the modulation frequency, a more detailed analysis shows that there is always some output either at the modulation frequency or at its second harmonic, so that the process may serve as a threshold detector. This may, for instance, form the basis of an intruder alarm actuated by pressure on the otpical fibre. A second, more subtle observation, born out by experimental evidence [7.32], is that if the output at the modulation frequency is averaged over a sufficiently long period, then the average output is a predictable function of the modulation depth. Thus the system may be linearised. The observation time for the averaging process is simply that time required for the values of the relative phase angle ϕ_{12} to have swept through the range $0-2\pi$. This may be artificially reduced by mechanically perturbing the system.

The attraction of the Fibredyne interferometer is that it is optically very simple. Coherent light is launched into the fibre, and the detector is placed at the other end to intercept about 25% of the output light. The requirement for precisely aligned reference and signal beams is completely eliminated. In effect, the reference and signal beams are different modes propagating along the same optical fibre.

Optical fibre interferometers offer the potential and the fact of extremely sensitive sensing devices. In acoustic detection, magnetometry and accelerometry, fibre interferometers have demonstrated performance levels comparable with, or exceeding, those attainable using other technologies. This comes about from the combination of the availability of coherent sources, the development of shot-noise-limited phase detectors and the use of appropriate materials in the 'measurand to strain' convertor alluded to earlier. Thus the optical fibre magetometer appears to offer sensitivities comparable with the SQUID device, and without cryogenics. The interactions are wide ranging, and it has even been proposed that gravitational telescopes could be fabricated using fibre interferometer technology [7.33]. Perhaps paradoxically, the principal difficulty with the all-fibre interferometric sensor lies, in many cases, in this extremely high sensitivity. The elimination of cross-sensitivities — for instance, variations in temperature affecting the validity of a strain measurement — proves to be somewhat difficult, and practical cross-sensitivity considerations limit the achievable effective sensitivities depending on the anticipated environmental conditions. Useful sensors are, however, available for a wide variety of applications.

7.4 Optical fibre phase sensors for mechanical variables

The most common form of optical fibre phase sensor is shown in Fig. 7.6. The basic silica fibre is usually coated with a medium which transforms the applied field into a longitudinal strain. This may be in the form of a compliant plastic to enhance acoustic sensitivity, a relatively stiff material — for instance a metallic coating — to reduce acoustic sensitivity, or a magnetostrictive material to couple the phase modulation to magnetic field variations. Typically the transforming coating will be drawn on to the fibre during manufacture, but the simple silica fibre may be attached to, for instance, a magnetostrictive strip or cylinder for magnetometer applications.

The interferometric optical fibre hydrophone has been the subject of a considerable research effort [7.27]. There are two principal motivations – the interferometric device may be made extremely sensitive, and the fibre may be simply configured in an endless variety of physical shapes. The polar and frequency response of an optical fibre hydrophone may therefore be readily tailored to a specific requirement. In the longer term, there is also the possibility of a passive multiplexed optical fibre transducer array.

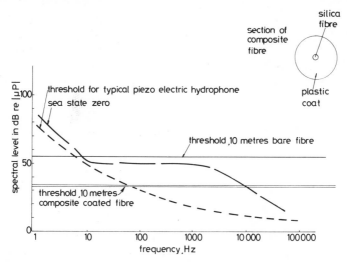

Fig. 7.8   *Theoretical pressure sensitivity of a single-mode optical hydrophone using bare silica, and using the composite structure as a strain transformer.*

The pressure sensitivity of bare silica fibre is shown in Fig. 7.8. At long interaction lengths this proves to be more than adequate for hydrophone applications, but there are considerable advantages to using a short sensor. Attenuation is reduced, and the difficulties involved with source phase noise in interferometers of different length arms are minimised. Also shown in Fig. 7.8 is a calculated sensitivity level for a fibre coated with a Teflon coating of 0·7 mm total diameter, in a composite structure similar to that shown in Fig. 7.8. The details of these calculations may be found in References 7.35 and 7.36, and detailed results are available in Reference 7.27 and the additional references cited therein. The important observation is simply that the sensitivity can be controlled by varying the thickness, the Young's modulus and the bulk modulus of the material coating the fibre.

This procedure modifies the intrinsic pressure sensitivity of the fibre. It is, however, useful only at relatively low frequencies, and the situation changes considerably when the dimensions of the overall structure become comparable with the acoustic wavelength. There is the additional factor that at high frequencies the conversion of pressure wave to longitudinal strain modulation decreases in efficiency. The use of coated fibres is therefore restricted to frequencies of less than about

100 kHz. The dynamic mechanical properties of the plastics involved are also not well characterised at higher frequencies.

There is, therefore, a range of frequencies – from about 100 kHz to about 10 MHz – where the silica fibre itself may form the basis of a simple pressure sensor. This principle has been exploited as a phase modulated data transmission highway [7.31], and has again proved to be a remarkably sensitive system. Experimental data have demonstrated that ultrasonic drive levels of considerably below one milliwatt to the optical fibre are more than adequate for the transmission of data at a few kilobits per second. At still higher frequencies the acoustic wavelength becomes comparable to the waveguide dimensions, and radial acoustic core modes may be set up. This results in the differential phase term increasing dramatically [7.13] in multimode fibre, since both compression and rarefaction are occurring within the fibre simultaneously. Accordingly the differential phase modulation becomes of the order of twice the peak modulation. This situation is shown diagrammatically in Fig. 7.9. In single-mode fibre the same situation will lead to induced birefringence, so that a sensor which is essentially a modulated retardation plate is feasible (see also Chapter 10).

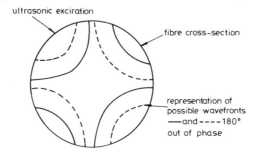

Fig. 7.9 *Radial stress distributions for a case when the acoustic wavelength is comparable to the fibre dimensions. Stress-induced birefringence is produced.*

The geometrical flexibility of the fibre hydrophone is perhaps its most useful feature. Fig. 7.10 shows some of the many possible formats. A simple fibre coil will form an omnidirectional sensor, provided that the coil diameter is much less than an acoustic wavelength. At higher frequencies, acoustic ring modes will propagate around the coil, resulting in a much modified fibre strain distribution [7.37]. At resonance, the measured sensitivity will be typically an order of magnitude over the low-frequency value. Two such coils may be assembled, one as the reference arm and one as the signal arm of an interferometer. The output is then proportional to the pressure difference between the two coils (again at low frequencies), so that this forms a pressure gradient hydrophone. A single coil may be made highly directional by winding on a long thin former, of diameter much less than the acoustic wavelength and length several wavelengths. The polar pattern of this directional element may be controlled by modulating the winding density along the former. Arrays of sensors are also clearly feasible.

Optical fibre accelerometers may be realised using similar principles. A freely mounted weight may exert an acceleration-dependent pressure on a fibre coil, or may be suspended by fibres each of which form the arms of an interferometer (see Fig. 7.11). The sensitivity and frequency response of these devices depend on the suspended mass and on the mechanical mounting configuration.

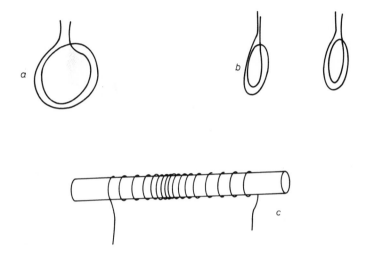

Fig. 7.10 *Some geometries feasible with the same optical fibre hydrophone technology. (a) Simple ring sensor (single-coil hydrophone) (b) pressure gradient sensor (two coils) and (c) a tapered directional sensor (shaded coil winding)*

Magnetometers using fibre interferometers involve a magnetostrictive material as the strain transformer. This may be a ferromagnetic metal — iron, nickel and cobalt, and alloys and compounds thereof. These materials all exhibit some hysterisis; all are sensitive to heat treatment effects and some to corrosion. High magnetostrictive constants are available — even from a polycrystalline form [7.38] — and the interaction may take place either through a fibre coating or via the use of a magnetostrictive mandrel. Metallic glasses may also be used as the magnetostrictive device [7.27]. These have the advantage that they may be drawn into convenient shapes, and may even be formed directly on the fibre.

The sensitivity of these devices has been measured and is very high (approximately 10^{-9} gauss metre; thus one kilometre is capable of resolving 10^{-12} gauss). There are also considerations similar to those of the hydrophone for the assembly of sensors of various frequency responses, and, in particular, gradient sensors for the measurement of magnetic dipoles.

Without exception, the optical fibre interferometer is a remarkably sensitive means of detecting environmental variations. However, the systems described here are only suitable for measurement of alternating fields — from frequencies of a few hertz upwards — and in the all-fibre form the phase detection process remains a compromise. There is considerable potential for both static measurements and improved phase detection processes, some of which are described in Chapter 13. It should also be noted that these interferometric sensors are essentially local

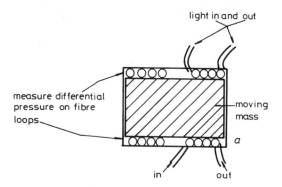

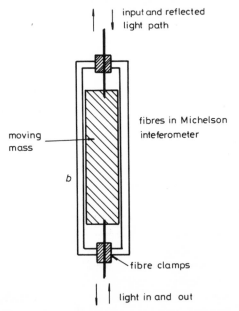

Fig. 7.11   *Optical fibre accelerometer configurations. The two fibres at each end of the mass would be in opposite arms of an interferometer. (a) Coils form 'reference' and 'signal' arms in all-fibre Mach Zehnder interferometer (b) fibres in Michelson interferometer*

sensors, in that the overall optical arrangement is unsuitable for transmission of data along the fibres used as the sensor. This again is discussed in more detail in Chapter 13.

7.5 The optical fibre Sagnac interferometer

The optical fibre Sagnac interferometer forms the basis of the optical fibre gyro-scope. The essential features of the system are shown in Fig. 7.12. The effect was first described by Sagnac in 1913 [7.39]. His original interferometer was an air path device with a phase detection resolution of approximately half a fringe (visual detection) and he obtained a sensitivity of approximately two revolutions per

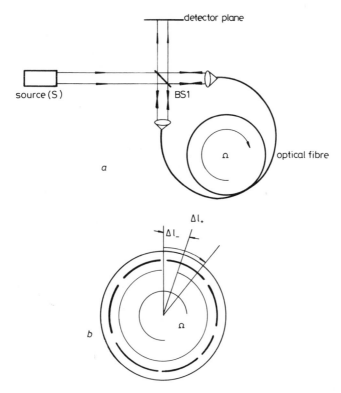

Fig. 7.12 *The optical fibre Sagnac interferometer (a) Interferometer configuration, and (b) the principles of the Sagnac effect*

second on about one square metre area. The Michelson and Gale experiment in 1925 [7.40] used this principle to measure Earth rotation. The optical fibre equiva-lent device increases sensitivity by several orders of magnitude, first by increasing the effective area of the interferometer loop by using several turns of fibre, and secondly by using electronic detection techniques. Inertial navigational perfor-mance is possible, in principle and in practice [7.41]. The basic mode of operation of the Sagnac interferometer may be understood by noting that if the loop for fibre in Fig. 7.12 is rotating at some angular velocity Ω, then light which is simul-taneously injected into the two input ends of the fibre will emerge at slightly different times depending on whether the light travels with or against the rotation. This results in an optical phase difference between the two propagation directions which is [7.42]

$$\phi = \frac{4\pi LR\,\Omega}{\lambda_0 c} \qquad\qquad (7.22)$$

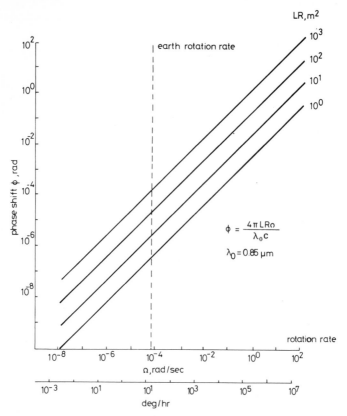

Fig. 7.13   *The phase shifts for a range of rotation rates for circular geometry optical fibre gyroscopes with the LR product as a parameter. A typical gyroscope will have an LR product between 10 and 100 m²*

where L is the total length of fibre wound on a cylindrical former of radius R. The values of this phase delay as a function of the parameter LR are shown in Fig. 7.13 for a source of wavelength 850 nm. These phase differences are in the microradian range for rotation rates of interest. The lengths of fibre involved are in the region of several hundred metres. There are therefore very large changes in the absolute phase of the fibre path with variations in temperature, pressure etc. The essential factor in the fibre optic gyroscope is that the reference and signal paths of the interferometer are exactly the same in both the fibre and any air gaps, so that, in principle, the only difference is due to the Sagnac phase shift. Maintaining this condition – reciprocity – is the principal concern in the design of a fibre optic gyroscope. The tolerance required on reciprocity is extremely high: the optical phase in the fibre path is typically 10^{10} radians, so that detection of a Sagnac phase shift of 10^{-7} radians implies that any other nonreciprocal effect should be kept well below 1 part in 10^{17}.

The condition for reciprocity may be stated very simply [7.43]. The fibre, which is single moded, may be viewed as a four-port network (see Fig. 7.14).

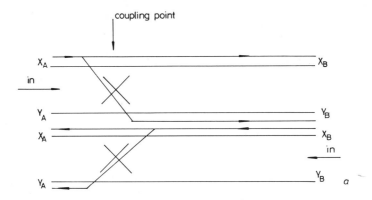

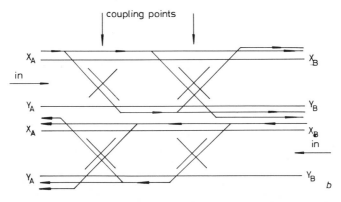

Fig. 7.14 *Illustrating polarisation coupling in a single-mode optical fibre, to demonstrate that reciprocity does not hold unless the input and output modes are exactly identical. (a) Single coupling point (b) two coupling points*

Lorentz reciprocity states that for a *linear* medium within the network, the network will be reciprocal if the input and output ports are taken through identical spatial modes. The definition of a spatial mode includes both the polarisation of the mode and its spatial distribution. Meeting this condition will ensure reciprocity if the fibre path is linear. This leads to the use of a polariser in the output arm of the Sagnac interferometer and a single-mode fibre section as a spatial filter. The ray paths involved in the interference process must traverse *exactly* the same paths, so that a second beam splitter is required to access the reciprocal port of the system. This leads to a 'minimum configuration' gyroscope (Fig. 7.15) which incorporates all the necessary features — the use of the reciprocal port and the addition of the polarisation and spatial filters.

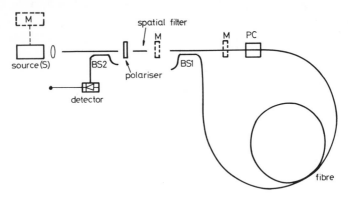

Fig. 7.15   *The minimum configuration gyroscope*

This ensures reciprocity, to the extent dictated by the rejection ratio of the polarisers and spatial filters, for a linear transmission medium. In practice the rejection ratios of a fibre polariser [7.44] or even a suitably oriented crystal polariser [7.42] can be adequate provided that a single spatial mode filter is also used. The latter component is readily formed from the fibre components themselves. But there remains the question of linearity of the transmission medium. It is tempting to argue that the power levels are so low (less than 1 milliwatt in each direction) that induced nonlinearities should be negligible. However, to tolerance levels of parts in 10^{17}, this is in practice not the case. The optical Kerr effect is a second-order nonlinearity in the electric constant of the material. Thus the electric polarisation P is related to the electric field through:

$$P = \epsilon_0 [\epsilon_r E + \eta E^2] \tag{7.23}$$

where η is the Kerr coefficient. This nonlinearity occurs in all materials, but is usually negligible except at very large electric fields [7.45]. However, it can be shown that the effect is not negligible for the fibre gyroscope [7.46], and the Kerr rotation rate error Ω_K may be shown to be [7.47]

$$\Omega_K = \alpha(1 - 2K)\langle I_0^2(t)\rangle) - 2\langle I_0(t)^2\rangle \tag{7.24}$$

where α is a constant, K is the coupling fraction of the coupler used to launch the light into the fibre, and $I_0(t)$ is the intensity of light launched into the fibre as a function of time. The averages take over the loop transit time and, finally, over the receiver integration time — the loop transit time is the order of tens of microseconds, the integration time may be of the order of a second. The Kerr-induced rotation error disappears either if a coupling coefficients are exactly 0·5 or if the source is modulated such that the difference of the weighted intensity moments averages to zero. Experimental evidence suggests that for an imbalance in K approaching 0·95 to 0·05, the induced rotation rate error is equivalent to about 20°/ hour. This is some four orders of magnitude above the sensitivity of this instrument (7.41] so that Kerr effect correction to about one part in 10^4 is required. One possibility is to modulate the laser source in a square wave, as suggested in Reference [7.47]. This appears to satisfy the requirements of the intensity relationship, but on closer examination it must be appreciated that the Kerr effect is instantaneous; it is therefore the mean of the intensity of the time-varying optical waveform, measured at optical frequencies, which concerns us, and not the low-frequency 'smoothed' envelope. The high-frequency waveform is the sum of the optical field emitted from the source, which may be a superluminescent diode, a multimode laser or a single-mode laser. Closer examination of the statistics of the waveforms from a superluminescent diode indicate that this source satisfies the intensity condition exactly, and a multimode semiconductor laser closely approaches the necessary requirements. Hence, with the appropriate optical source, the Kerr nonlinearity may be removed [7.48].

Thus, meeting the exact reciprocity condition is a question of providing the correct optical path and the correct optical source. There are alternative approaches, which all centre around measuring the nonreciprocity and filtering this out using electronic processing. This offers some potential for gyroscopes with rate performance of 100° hour, but for inertial grade systems (0·01°/hour) the reciprocity requirements must be met.

Reciprocity is one of the key issues with the optical fibre gyroscope. The other is the detection of the phase modulation introduced by the rotation. The Sagnac interferometer is a true zero path difference interferometer, so that detection of very small phase shifts is fundamentally impossible unless the operating point of the interferometer is somehow shifted to the quadrature condition. The appropriate intensity versus relative phase graph is shown in Fig. 7.16. Possible operating points — A and B on the diagram — obviously require that the interferometer be arranged to be nonreciprocal. However, one possible way around this is to use an alternating bias point, perhaps simplest to conceive as switching between the two states A and B. A small Sagnac phase shift $\Delta\phi$ will then produce a square-wave output, whose amplitude is dependent upon the Sagnac phase and whose phase (zero or 180°) gives the sign of the rotation. Similar results may also be obtained using a sinusoidal phase variation, and this may be implemented by using a piezoelectric cylindrical phase modulator at one end of the fibre loop. The driving frequency to the modulator is arranged so that the relative phases of the two beams are inverted during the transit time. The output from the detector at the fundamental frequency (the phase modulation frequency) is then found to have an amplitude proportional to the $J_1(\phi_{sag})$ and a phase relationship which determines the sign of the rotation.

This is one method by which a nonreciprocal phase shift may be applied to the Sagnac interferometer to facilitate the detection process. This forms the basis of the majority of detection schemes used in fibre gyroscopes. The average nonreciprocity is zero, since the sign changes many times during the integration time of the detection electronics. There are numerous other phase detection schemes in use, including heterodyne interferometric gyroscopes [7.49] and phase nulling gyroscopes

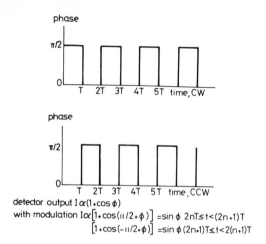

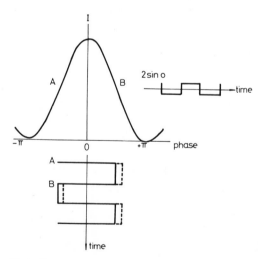

Fig. 7.16   *Phase detection in fibre gyroscopes*

[7.50]. One version of the phase nulling gyroscope [7.51] has attained approaching shot-noise-limited performance, but the overall optical configuration is only suited for laboratory use. The detection limits on other versions of the gyroscope using the same principles are well above this limit. The motivation for examining other techniques is that there is a considerable potential for these concepts in practical situations to overcome some drift and thermal effects inherent in the use of PZT devices, and in relying on the stability of the source wavelength over an extended period.

Noise in fibre optic gyroscopes arises from a number of sources. There is the normal optical detection noise — shot noise and thermal noise in load resistances, along with multiplication noise if APD detectors prove to be necessary. There are also important secondary noise sources associated with the optical components, in particular $1/f$ detection noise, which is unavoidable and requires the use of detection frequencies of over 100 kHz, and excess source noise, which depends on a complex combination of the source drive circuit noise and the inherent source characteristics. Semiconductor lasers are known to exhibit frequency-dependent excess noise, with peaks at low and high (gigahertz) frequencies. For a good laser diode, the noise level between 100 kHz and 100 MHz is approximately shot noise. The detection frequency is essentially determined by these noise considerations as well as availability of suitable modulation components.

Another inherent source of noise in the fibre gyroscope is backscattered radiation. The principal loss mechanism in an optical fibre is Rayleigh scatter. Thus radiation which is scattered from the core enters a full 4π solid angle, and a part of this is collected by the core of the optical fibre and transmitted back to the source. (The optical time domain reflectometer [7.52] exploits this effect to locate scattering centres.) The fraction of the scattered radiation which is directed to the source is readily calculated from the fibre numerical aperture *as viewed from within the fibre*. This is $(2\Delta n/n)^{1/2}$ and so the fraction guided is $(\Delta n/2n)^2$, which is typically 5×10^{-3}. For a guide with a loss of 10 dB/km, the power loss is approximately 0.1% per metre, so that the backscattered power returning to the source is 53 dB down on the source level from a one metre section of fibre. If this returned power is coherent with the interfering radiation from the two paths around the interferometer, then it can produce a phase error in the region of 10^{-3} radians. This exceeds the rotation rate signal for rates up to approximately Earth rotation (15° per hour). The effect is negligible if the backscattered radiation is not coherent with the source [7.53].

More detailed calculations [7.42] indicate that, for inertial quality performance, coherent backscatter can prove troublesome for coherence lengths exceeding a few microns.

Interferometer drift is essentially low-frequency noise. This can occur due to component drifts, especially in the launch coupler, resulting in variations in the induced Kerr effect for noncompensated sources. The majority of detection schemes currently in use depend on intensity stability of the source, and any intensity variations will result in an erroneous reading. Phase nulling and heterodyne techniques eliminate this as a source of error — intensity fluctuations cause only a change in noise levels in these systems and otherwise do not affect the perceived readings. The scale factor of the gyroscope also depends on the loop length and area, both of which are temperature sensitive, though these can be accurately computer corrected using thermal sensors. However, the dependence on optical wavelength is more critical. The temperature coefficient of wavelength of a semiconductor laser source is typically about 0·1% per degree. An inertial grade instrument will require scale factor stability of better than 1 in 10^5, so that clearly either some form of wavelength stabilisation is required or an entirely novel detection scheme should be devised to extract the information without relying on optical wavelength calibration. Wavelength stabilisation with semiconductor lasers is feasible to megahertz accuracies [7.54, 7.55], so that this would be one possible approach to scale factor stability. Other laser sources with less pronounced temperature coefficients may also be used, though many of these − for instance gas lasers − exhibit a high degree of coherence, and interferometers using this source are susceptible to coherent backscatter noise. It is likely that a combination of a different laser source and a novel detection scheme will be required to completely eliminate scale factor drift. This will, of course, have to be compatible with the other requirements of the system imposed by reciprocity and noise considerations.

The fibre gyroscope is a particularly interesting example of an optical fibre sensor system. It has a performance potential (already realised by its close relative the ring laser gyro [7.56]) which is comparable with all but the most precise of mechanical gyroscopes, but the fibre gyro has no moving parts and no plasma tube, so it should be inherently much more reliable than any system previously developed. The optical concepts in the fibre gyroscope also bring together the application of many of the fundamental principles of optics [7.57]. This short description has illustrated the main points, but there is a considerable and expanding literature available to cover the details.

7.6 Optical fibre interferometric sensors − discussion

Optical fibre interferometric sensors are among the most sensitive devices yet discovered. Their applications include industrial, military and scientific measurement. The majority of common measurands have been monitored using interferometric sensors, and a number of novel applications − for instance in photoacoustic spectroscopy [7.58] and in the measurement of acoustic surface waves [7.18] are also emerging.

Interferometric sensors as described in this chapter are limited to performing the sensing function only. Transmission using the same light is precluded in all these devices because of the possibility of further phase modulation on the transmission path. A number of techniques are evolving to utilise both the transmission and sensing properties of fibres simultaneously and separately. Some of these are discussed in further detail in Chapter 13, in which the essential problems are analysed and some of the solutions described.

Chapter 8

Frequency modulation in optical fibre sensors

8.1 Introduction

Frequency modulation of light occurs under a limited range of physical conditions. The principal one of interest, and the only one considered here in any detail, is Doppler shift of a beam reflected or scattered from a moving target. There are several other circumstances under which the frequency of light is shifted. These include the numerous absorption and phosphorescent phenomena – some of which are considered in the next chapter – and quantum mechanical interactions such as Brilloin and Raman scattering.

The Doppler effect is well known. If radiation at a frequency f is incident on a body moving at a velocity v as viewed by an observer, then the radiation reflected from the moving body appears to have a frequency f_1 where:

$$f_1 = \frac{f}{1 - v/c} \simeq f[1 + v/c] \tag{8.1}$$

There are countless examples of this in radar, optics and acoustics and the Mössbauer effect in nuclear physics. The simple basic treatment is readily extended to include the possibilities of source, observer and target all moving relative to each other. In an optical system, Doppler shifts provide a very sensitive detector of target motion. For instance, with an HeNe laser as the light source, the frequency shift is 1.6 MHz per metre per second. A laser Doppler probe should then be capable of detecting target velocities in the range from microns per second to perhaps 10–100 metres per second, depending on the choice of detection electronics [8.1].

Laser Doppler velocimetry is widely used in applications such as fluid flow measurement using argon laser sources and atmospheric turbulence measurements using carbon dioxide lasers operating at a wavelength of 10.6 microns [8.2]. The use of laser Doppler as the basis of an optical fibre probe has also been described in the literature [8.3, 8.4]. In this chapter, the fundamental features of a fibre optic Doppler probe are described, and an assessment of possible applications is given.

The basic signal to noise ratio calculation for a laser Doppler system may be calculated by assuming that the system is shot noise limited — which is usually the case — and that the frequency difference produced by target motion is measured with reference to a local oscillator beam of power P_{lo}. If the local oscillator is a wavelength λ, this corresponds to N photons per second where:

$$N = \frac{P_{lo}\lambda}{hc} \tag{8.2}$$

and the shot noise level in a 1 Hz bandwidth is then proportional to $\sqrt{N}$ (neglecting contributions from the finite quantum efficiency of the photodetector). The signal returned from the target power P_t mixes with this local oscillator beam to produce a difference frequency signal of power $2(P_t P_{lo})^{1/2}$ (this is readily shown by summing the fields at the detector, and assuming perfectly aligned beams for maximum fringe contrast). This gives a minimum detectable signal of $P_t \sim hc/\lambda$ in a 1 Hz bandwidth, corresponding to power levels of the order of 10^{-19} watts. Heterodyne detection systems with sensitivities of this order have been reported [8.5]. The threshold ratio of local oscillator power P_{lo} to target power P_t may also be derived from these considerations, and is found to be

$$\frac{P_t}{P_{lo}} > \frac{1}{N} \tag{8.3}$$

The derivation of the signal to noise ratio for a heterodyne detection system is considered in more detail in Reference 8.6.

The detection threshold derived above does, of course, assume perfect mixing between the reference beam and the target beam, and shot-noise-limited detection. Thermal noise and, for lower frequencies, $1/f$ noise may introduce a significant deterioration — though these contributions may be estimated using the techniques of chapter 5. Alignment effects are also important. It is essential for optimum performance that all the optical power falls on the detector, and that the interference produces only one fringe across the detector face. In many practical situations, it is difficult to achieve this alignment precision. For instance, with a typical small detector of 100 microns diameter, the interference signal drops off by 10 dB for a one micron wavelength optical signal if the interfering beams are misaligned by 7 milliradians. There is also a considerable art to the successful detection of return signals from laser Doppler interferometric systems and related devices. In particular, the return signals are frequently very small, and often contaminated with noise corresponding to random events which occur simultaneously with the events to be monitored. Photon correlation techniques have become well established [8.7] as the preferred means for signal processing in these systems. There are, of course, circumstances where the complexity of a full correlator is not required for successful signal interpretation — this needs to be assessed for the individual application.

8.2 Optical fibre Doppler systems

The principal advantage of an optical fibre laser Doppler probe lies in the fact that the position of the measuring zone may be adjusted without recourse to realignment of the system launch and receive optics. However, the fibre probe does introduce a number of characteristics peculiar to the fibre system.

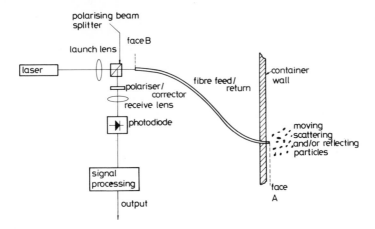

Fig. 8.1   *Schematic diagram of a fibre optic Doppler anenometer*

A schematic diagram of a fibre Doppler probe is shown in Fig. 8.1. A laser source is fed into a multimode optical fibre via a polarising beam splitter and launch optics. The other end of the fibre is immersed into a fluid in which the velocity of either the fluid or bodies within the fluid is to be measured. Light is scattered within this fluid, and some of this light is collected by the fibre and returned. The scattered light is randomly polarised, so that at the laser end of the fibre, the polarising beam splitter returns half of this scattered light to the photodetector.

There remains the problem of deriving a reference for the returned beam to interfere with. This beam must be derived from a point which is stationary with respect to the mobile bodies to be monitored, and the only point to meet this requirement is the end face A of the fibre. Thus the reference is usually derived by using the reflections from this end face. This reflection depends on the relative refractive indices of the fibre and the medium, and is always less than 4% in power (this is the total reflectance for a glass—air interface [8.8]). This is a lossy means to derive the reference signal but, with the important proviso that there are only very small interfering signals due to spurious reflections from the remainder of the system, this can prove to be adequate.

Spurious reflections occur primarily at the launch end of the fibre, but, on input, the polarisation of the laser is accurately defined by using a polarised laser and aligning this polarisation accurately with the polarising beam splitter. Thus, any reflections originating at the front end of the system will be largely polarised in the same direction as the laser source. They will therefore be reflected straight back into the laser. For an HeNe source this will have little effect, but there may be serious consequences for a semiconductor source. The multimode optical fibre very rapidly depolarises the input light — usually within a distance of a few centimetres — so that any return signal from the fibre, which includes the end reflectance and backscatter from along the fibre length, will have about half its intensity in a polarisation orthogonal to the input direction. Thus light from the fibre end face A has a component which is diverted by the polarising beam splitter on to the photodetector. Therefore, in principle, the reference beam component is derived from the correct plane. In practice there are small birefringences in the components at the launch end, especially lenses, and there are also the effects of the finite extinction ratio of the polarising components to take into account. This is typically no more than 60 dB. Additional polarising components may be used to improve upon this figure.

It is useful to compare the power levels returning to the detector from the various planes within the system, all referred to the laser input power:

> Launched power : 0 dB
> Power reflected from face A: 24 dB for fibre—water interface
> Power to detector from face A: 27 dB
> Power to detector from face B: 74 dB

this assumes 4% intensity end reflection and 60 dB polariser with no local bi-refringence. It should be noted that $5°$ birefringence could increase this signal to -35 dB.

This procedure establishes the reference, and the necessary precautions required to maximise the reference power and minimise interference with this reference from spurious reflections in the system. Careful attention to the polarisation properties of the launch end optics is clearly the important issue. However, the spurious signals are comparable in intensity to — and often exceed — the backscatter signal from the moving targets, so that there will be an additional velocity signal due to the motion of the spurious reflecting surfaces with respect to end A. How-ever, in most circumstances the velocities of the target will differ radically from the relative velocities of ends A and B of the fibre (typically due to thermal changes etc.), so that simple frequency filtering will be more than adequate to separate the two.

The returned signal from the targets depends on the intensity of backscattered radiation and attenuation within the medium, and the receive area and numerical aperture of the fibre. The physical process involved is shown schematically in Fig. 8.2. Attenuation in the medium takes place through both scattering and ab-sorption. The scattering component of the attenuation gives rise to some return signal. The fraction of the scattered radiation collected is determined by the accept-ance numerical aperture of the fibre and the core area. An estimate of the power returned may be made assuming that the fibre is a step index fibre and that the radiation from the fibre is of uniform power density in the emerging cone of light.

The power arriving at a plane at a distance z from the fibre end is then:

$$P_z = P_0 e^{-\alpha z} \tag{8.4}$$

where P_0 is the power launched into the medium from the fibre and is approximately the power launched by the laser (neglecting fibre transmission losses, launch losses and reflections. For a probe length of a few metres, these would total only a few decibels at most). α is the attenuation coefficient in nepers/metre.

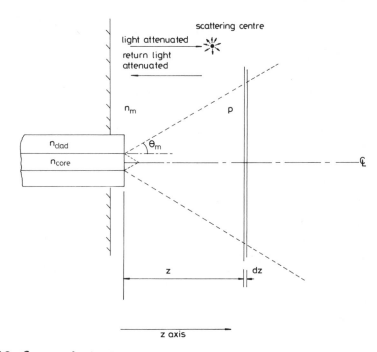

Fig. 8.2 *Geometry for describing the collection of light scattered from mobile particles*

The total power scattered from the element of length $\mathrm{d}z$ at z is:

$$P_{\text{scat}} = P_z e^{-\alpha_s \mathrm{d}z} \simeq P_z [\alpha_s \mathrm{d}z] \tag{8.5}$$

where α_s is the scattering attenuation coefficient. This is scattered uniformly throughout a solid angle of 4π. Following the treatment in Reference 8.3, the power coupled from this scattering plane back into the fibre may be estimated by considering the scattering plane as a Lambertian source coupled into the fibre, for which the coupling efficiency η is [8.9]:

$$\eta = \frac{A_f}{A_z} (NA)^2 \tag{8.6}$$

where A_f is the core area, A_z the source area at z and NA is the numerical aperture of the fibre, given by:

$$NA = \frac{(n_{clad}^2 - n_{core}^2)^{1/2}}{n_m} \tag{8.7}$$

for radiation into a medium of index n_m.

Thus, the power scattered back into the fibre from z is P_{rz}, where

$$P_{rz} = \frac{A_f}{A_z} P_{scat} e^{-\alpha z} (NA)^2 \tag{8.8}$$

Noting that the source area is $\pi(z \tan \theta_m + a)^2$ (where a is the core radius) gives for the total return power P_r:

$$P_r = \frac{A_f}{2} P_0 \int_0^\infty \frac{e^{-2\alpha z} \alpha_s dz}{\{z \tan \theta_m + a\}^2} \tag{8.9}$$

where the factor $\frac{1}{2}$ accounts for the fact that half the scattered power is back-scatter and the other half forward scatter. The total attenuation is α and the scattering component α_s in nepers/metre. This expression may be manipulated into the form:

$$\frac{P_r}{P_0} = \left\{ \frac{a - NA}{\tan \theta_m} \right\}^2 \alpha_s \alpha e^{2\alpha a / \tan \theta_m} \int_L^\infty \frac{e^{-x} dx}{x^2} \tag{8.10}$$

where x is a dummy variable obtained by suitable substitution in eqn. 8.8, and $\theta_m = \sin^{-1}(NA)$. For the limiting case of a small numerical aperture, that is $NA \sin \theta_m \simeq \tan \theta_m \simeq \theta_m$:

$$\frac{P_r}{P_0} = a^2 \alpha_s \alpha e^{2\alpha a / \tan \theta_m} \int_L^\infty \frac{e^{-x} dx}{x^2} \tag{8.11}$$

with the lower limit of the integration L given by:

$$L = 2\alpha a / \tan \theta_m \tag{8.12}$$

Some further manipulation of these expressions (see Reference 8.3) gives the expression:

$$\frac{P_r}{P_0} = \frac{(NA)^2}{2} R F(L) \tag{8.13}$$

where R is the ratio α_s / α and the function

$$F(L) = L(1 + Le^L E_1(L)) \tag{8.14}$$

is plotted in Fig. 8.3. Here:

$$E_1(L) = \int_L^\infty \frac{e^{-x}}{x} dx \tag{8.15}$$

With this information, estimates may be made for the detection threshold of the system as a function of the total attenuation of the medium.

As an example, consider the case where P_0 is one milliwatt at a wavelength of 633 nm (helium neon). This would be typical for a system powered by a laser emitting 3 to 5 milliwatts total. The local oscillator power would then be $- 27$ dBm (or 2 microwatts) corresponding to $6\cdot38 \times 10^{12}$ photons per second. This sets the minimum value of P_r/P_0 as $1\cdot6 \times 10^{-13}$, or an overall maximum return loss of 128 dB.

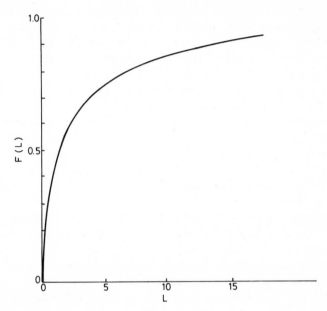

Fig. 8.3   *The function F(L). See text for description of the use of this graph*

This assumes that the overall mixing process is perfect and that other sources of noise, most notably $1/f$ noise, do not contribute. In estimating the requirements on attenuation, we also need to include the loss in the polariser, which is 3 dB, losses due to reflections, which will be at most 1 dB in a well designed system, and a mixing loss due to misalignment of the reference and signal beams. It is particularly difficult to estimate this in detail, since multimode fibre is involved, and the interference process takes place between the modal spectrum reflected at plane A in Fig. 8.1 and the modal spectrum launched by the scattering process. The latter will probably be fully filled, but it is far from clear what the former will be. On a purely arbitrary basis, it is probably safe to add 20 dB for mixing loss. In a single-mode system this would be high, but here it represents a reasonable starting-point. This then gives a minimum detectable return loss of 104 dB, so that P_r/P_0 must exceed 4×10^{-11}.

A typical optical fibre for use in this application may have a core diameter of

50 microns and a numerical aperture of 0·15 in air, which corresponds to 0·15/ 1·33 = 0·113 in water. If we assume that the attenuation in the medium is entirely due to scattering then $R = 1$, and

$$\frac{P_r}{P_0} = 0.0064 \, F(L)$$

This gives the minimum detectable value of $F(L)$ as approximately 10^{-8}. For very small L, $F(L) \simeq L$, so that the detection threshold occurs when the scattering attenuation exceeds 2×10^{-5} nepers/metre. This estimate demonstrates that, provided there is some scattering attenuation in the medium, then, for the 1 Hz bandwidth case considered, most media will return a detectable signal. This is borne out by the experimental observation that significant signals can be detected from ordinary tap water, but not from distilled or filtered water.

Thus the sensitivity of the measuring technique is well established. These approximate estimates of performance are confirmed in practice. However, one other parameter requires discussion, which is the volume over which the measurement actually takes place. Returning to Fig. 8.2, one can readily demonstrate that for scatterers at a distance exceeding $a/\tan \theta_m$ from the end of the fibre, the light coupled back into the fibre end drops rapidly. Thus the maximum penetration distance into the medium is of the order of a few core radii. A more detailed analysis (presented in Reference 8.3) confirms that this is generally true, and for heavily attenuating media the penetration depth is considerably less, saturating at approximately two core radii. Thus this probe will measure particle velocity in fluids to a distance of, at most, a few millimetres from the fibre end.

The overall characteristics of the basic fibre optic Doppler probe are then that it is a very sensitive detector for the motion of scattering bodies in a transparent medium. However, its penetration depth is limited to a few millimetres, so that it is difficult to accurately monitor fluid flow with this system in all except a few special circumstances, since the presence of the probe will disturb the flow patterns over this range. However, it is readily suited to the detection of moving targets or to the detection of mobile bodies in suspension. The probe will detect velocities as low as one micron per second, and up to metres per second or above depending on the detection electronics, corresponding to frequency offsets ranging from a few hertz to tens of megahertz.

8.3 Development of the basic concepts

There are a number of obvious areas into which the basic concepts of the fibre Doppler probe can be extended. Perhaps one of the principal limitations of the basic system is the penetration range into the medium to be monitored. However, once it is appreciated that the limitation is in effect imposed by the area of the illuminating fibre, and the numerical aperture into which it is launched, it is fairly simple to conceive of optics which will manipulate these parameters. In fact a simple lens structure, shown in Fig. 8.4a, produces for $d < f$ a virtual image of the source at a position u from the lens given by the simple lens formula (see Appendix 1), with an image radius $r_i = au/d$ and image numerical aperture $NA_i = NAd/u$ (a is the core radius). The range of accepted backscatter — neglecting the effects of attenuation in the medium — is then of the order of r_i/NA_i, and is extended by a factor of the order of the square of the magnification of the image. This procedure is only useful when attenuation within the estimated range is small — less than a few decibels. Otherwise, the bare probe will be more effective. The effect of this procedure on the returned optical power may be estimated using the fact that $F(L)$ is, for small L, approximately L, and that the function saturates for larger L.

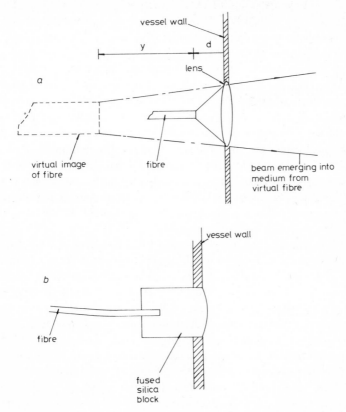

Fig. 8.4 *The use of lens structures to modify the collection geometry in low-attenuation media*

Thus, for the returned power ratio with a given fibre and medium, we may put, for media with small attenuation — that is, with small L —

$$\frac{P_r}{P_0} \sim \frac{NA^2}{2} RL = NA R 2\alpha a \tag{8.16}$$

The optical system maintains NAa as a constant, so that for small L the fraction of power returned remains constant as the magnification of the fibre end is increased, but at higher magnification the returned power drops. As an example, a heavily attenuating medium with an attenuation of 100 dB/m has an L value of approximately 0·06 for the fibre discussed in the previous section. The function $F(L)$ is almost linear up to $L = 0·03$, so that magnifying the fibre by a factor of two increases the penetration by a factor of four, but maintains the returned power levels at an approximate constant. Note also that the returned power level is in the region of $10^{-4}P_0$, and excellent signal to noise ratios are therefore available.

The simple lens structure proposed here as a means of extending the beam penetration introduces extra reflecting surfaces, and will distort the reference signal. It is probably preferable to use a structure as in Fig. 8.4b to perform an identical function in a slightly different manner. However, with careful design, and the use of appropriate media in which scattering attenuation is significant, penetrations of the order of metres appear to be feasible.

There are numerous other refinements available. The problems due to spurious reflections, especially from launch optics, are minimised by the use of polarisation selectivity, but due to the relatively low reference signal and the limits to polariser rejection ratios, it may be desirable to reduce these signals still further. This is simply effected by ensuring that the launch reflections are not coherent with the reference signal. In this case, they will simply add to the shot noise – by a minute amount – but will not interfere with the returned signal. This implies that there is a maximum desirable coherence length for the laser source used in these systems, and that this coherence length is approximately the penetration depth into the medium. This could be an advantage for semiconductor laser sources, though this has to be weighed against their higher phase noise levels.

The use of a reference beam at the same nominal frequency as the signal beam is simple to implement, but a frequency-shifted (heterodyne) reference beam is often preferable. There are two principal reasons. A heterodyne reference shifts all the desirable information in frequency to higher than 10–100 kHz, thereby avoiding problems with $1/f$ noise in detectors. Secondly, the heterodyne reference allows the system to detect both direction and speed; the homodyne system determines speed only. Generating the required frequency-shifted optical signal is simple; Bragg cells are the most convenient. However, it is far from easy to manipulate this beam so that it originates from the correct reference plane – the interface between the fibre end and the medium – and to ensure that the beam is spatially coherent with the signal beam (for optimum mixing efficiency) but has the appropriate temporal coherence to avoid false signals due to spurious reflections. There have been occasional valiant attempts [8.10] but none has, to date, provided a satisfactory solution.

8.4 Discussion

The fibre optic Doppler probe is, in its basic form, a simple instrument suitable for use in probing a volume close to the fibre end – within at most a few millimetres – where it can examine phenomena such as Brownian motion and the movements of micro-organisms. In this form it is probably unsuited to a flow measurement role, as for instance in the laser Doppler flowmeter, since the beams do not penetrate adequately into the medium. Penetration may be increased by using an imaging system. The use of a multimode fibre to transfer the beam from the laser source to the measurement region, however, destroys the spatial coherence of the source, and the beam is thus not suited to interferometric measurements which would be directly analogous to those used in conventional laser Doppler velocimeters. Single-mode fibre could be incorporated into the system, along with collimating optics at the probe, and an exact analogue of laser Doppler probes could be built. A two-fibre or a single birefringent fibre system could provide one way to introduce a heterodyne signal. There is one other contrast with conventional velocimeters: the optical power levels in a fibre system would be much lower, since about a hundred milliwatts is the maximum which can be carried in a single-mode fibre without detrimental non-linear effects.

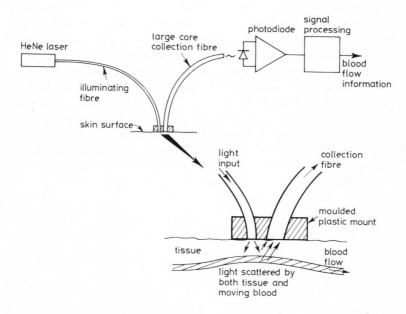

Fig. 8.5   *Principles of an optical fibre blood flow Doppler probe*

Finally, it should be noted that some success has been achieved with fibre flow-meters using time of flight techniques and correlation processing [9.11]. Though this is not an FM system, it is closely related in final applications. The principle is very simple: the optics produces two spatially separated images of the source in the moving medium, then images these two images on to the end of two optical fibres. As particles flow in the fluid, through first one image and then the second, they modulate the light intensity. The modulations in the two images are highly correlated when the time separation of the two corresponds to the velocity of the fluid. Thus a correlation measurement gives the flow rate. It would be a logical extension to this technique to incorporate Doppler processing.

There are numerous other approaches to the design of a fibre optic Doppler flow-meter. They are all very similar to the device described in detail in this chapter, but the derivation of the reference signal — that is, the unmodulated optical signal from a suitable stationary plane — is implemented in different ways. For instance, a device which has found many practical uses in medical applications is shown schematically in Fig. 8.5. Here, the returned signals are transmitted via a separate fibre from the source signals. The return fibre is a short, large-aperture (perhaps one millimetre in diameter) high NA fibre. This collects scattered light from the target, involving both light from stationary tissue in the subject and from the mobile blood. Spectral analysis is then readily applied as previously. This arrangement also allows for some simplifications in the signal processing, and the assembly is described in full detail in References 8.12 and 8.13.

Frequency modulation in optical fibre sensors may only be caused by some form of Doppler shift, unless nonlinear effects are occurring. But stimulated Brioullin scattering and Raman scattering are unimportant at power levels under a hundred milliwatts or thereabouts, depending on the interaction length in a single-mode fibre. It is feasible to configure a sensor system based on nonlinear effects in fibres, but the high power levels render the concepts relatively unattractive. Not only is the optical source now a much more involved and expensive affiar, but also the system safety is considerably impaired, from the aspects both of eye damage and of explosion risks in hazardous areas. Perhaps some significant materials develop-ments will change the situation [8.14]. There is however, considerable potential for Doppler modulated systems in the measurement of flow and motility. This is one area which will see a modest but continuing expansion in its applications.

Chapter 9

Wavelength distribution (colour) sensors

9.1 Introduction

There are numerous physical phenomena which influence the variation of reflected or transmitted light intensity with wavelength (that is, perceived colour). Perhaps the best known of these are the chemical indicator used for monitoring pH etc., and pyrometers used for temperature measurement. In fibre optic colour probes, the fibre simply serves to feed light to the monitoring region and return the modulated light for analysis.

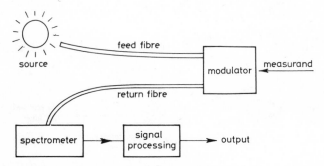

Fig. 9.1 *The principal feature of a colour modulation sensor*

The principal components in a colour probe are shown in Fig. 9.1. The source of light will be typically in the visible or ultraviolet, and the returned illumination in the infra-red to visible regions. The attenuation characteristics of the fibre, even in very short lengths, can modify the feed and return spectra considerably at these relatively short wavelengths. This implies that the calibration procedure for the instrument will be a function of fibre length and type, but it is often the case that the transfer characteristic of the fibre will remain constant with aging, temperature etc., so that the calibration, once implemented, will remain.

The critical components are then the source and the spectrometer, and it is the properties of these which play the most important role in determining the overall stability and resolution of the sensor system. In most colour modulation systems, the optical source is an incandescent lamp or mercury arc. The spectral properties of the former in particular are far from stable, so that it is usual to build into the sensor some form of continuous technique for recalibrating the system. Similarly, the spectrometer is often also unstable — with thermal variations in particular — and the calibration procedure can also compensate for this. Finally, as a general rule, ratiometric measurements are usually taken, so that the parameter measured is colour as determined by optical intensities at, at least, two different wavelengths.

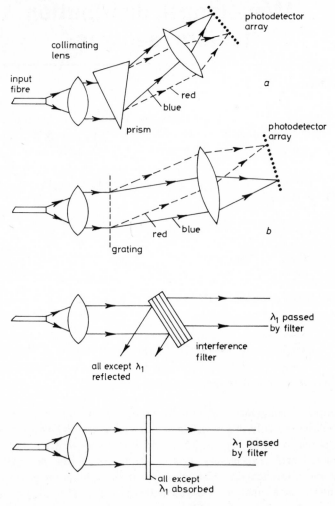

Fig. 9.2   *The basic principles of the most common spectrometer components*

There are two other limitations imposed on colour sensors by the choice of source. These are that the transmission range is usually fairly short — typically metres, though possibly up to 100 metres — and that the type of fibre used is invariably large-core, high-numerical-aperture material for efficient coupling of a spatially incoherent source. These are generally high-loss fibres, though they are extremely easy to handle.

Spectrometers can vary in complexity over a wide range. The simplest, used in an optical fibre pyrometric probe [9.1], is to exploit the wavelength sensitivity of a simple single photodetector. As the wavelength distribution of the light incident on this detector is varied, the detector current will also vary, provided that there is an adequate match between the detector and the phenomenon to be monitored. Prism and grating spectrometers are also available as standard equipment which is readily modified for use in a fibre colour probe. Colour filters, based either on interference filters or on absorption in various dyes, are also readily available. There is a fundamental difference between dye-based (absorption) filters and interference filters in that the interference filter reflects radiation which it does not transmit. This is shown diagrammatically in Fig. 9.2.

The spectrometer resolution required in colour modulation devices is usually fairly low, so simple wavelength measurement techniques are adequate. The resolution of a prism spectrometer depends on the dispersion in the prism, the width and divergence of the beam and the focal length of the final lens which focuses the analysed light on to the detector array. The resolution of the grating spectrometer depends on the number of lines of the grating intercepting the input beam, and that of interference filters depends on the number of layers in the device, the refractive index difference between the layers and the accuracy with which the device has been fabricated. A detailed treatment of the tools of spectroscopy is beyond the scope of the present discussion, but the principles are described in detail in Reference 9.2.

Highly stable measurements of the relative intensity of the spectral components are difficult to achieve in practice. The reason is simply that, in a photodetector array, the relative response of the detectors at different wavelengths is a function of temperature and aging. This implies the necessity for calibration of the detector array against a known colour each time the instrument is used, or the use of a single detector which is multiplexed between the various spectral components of interest. Examples of both approaches are described below.

9.2 Techniques for colour modulation

There are four principal areas in which colour modulation may be exploited. These are in chemical analysis using indicator solutions, in analysis of phosphorescence and luminescence, in the analysis of black body radiation, and in the use of Fabry Perot, Lyot (polarisation based) or similar optical filters in which the transmission characteristics of the filter are made to be a function of an external physical parameter. Fibre optic probes designed to monitor all these parameters have been evaluated.

The fibre optic pH probe shown in Fig. 9.3 is typical of applications requiring remote chemical analysis. The indicator, in this case phenol red, is used to dye small polyacrimide spheres of diameters in the range 5–10 microns, and these are contained within a permeable membrane. The indicator has a transparency which is a very sensitive function of pH in the red, but is virtually independent of pH in the green. Thus, measuring the ratio of green light transmitted to red light transmitted is a direct indicator of the pH of the solution in which the sensor head is immersed, provided of course that the colour temperature of the tungsten lamp remains reasonably constant [9.3]. The ratio is measured in this system by the very simple technique of using two absorption filters, one red and one green, in a rotating wheel, thereby switching the light on one detector between the two colours. The ratio of the signal during the red period to that during the green period is then related uniquely to the pH. The ratio may be derived by measuring the mean level and the amplitude of the square wave produced while the wheel is in motion.

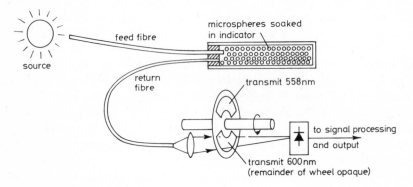

Fig. 9.3   *An optical fibre pH probe*

This system requires precalibration, since the relative sensitivity of the detector at the red and green wavelengths varies with temperature and aging, and the colour temperature of the light will drift. It is also preferable to use a stable DC source to power the lamp, though in principle the AC variations in light output may be smoothed in the signal processing and by proper choice of the rotation rate of the chopping wheel. The probe described in Reference 9.3 was designed for pH measurements of blood, and proved to have a resolution and repeatability of 0·01 in a pH range of 7 to 7·4. The values also agreed with those obtained by standard techniques. The principles can clearly be applied to a wide range of analysis problems involving indicator materials.

The same basic methods may also be applied to temperature measurement. Fig. 9.4 shows an overall schematic diagram of a fibre optic temperature probe based on the variation in the phosphorescence spectrum of a rare earth phosphor $((Gd_{0.99}Eu_{0.11})O_2S)$. The phosphor is illuminated in the ultraviolet, and emits a temperature-dependent spectrum (Fig. 9.5) in which the intensity of the red 'a' line increases with temperature whereas that of the green 'c' line decreases. The ratio of the two is a single-valued function of temperature, and is largely independent of the excitation spectrum since both lines are excited by the same part of the illuminating spectrum.

Measurement of the ratio is effected using the optical arrangement in Fig. 9.4, which utilises interference filters as the spectral analysis technique. In this case the two spectral components of interest are measured on different photodiodes, so that precalibration to correct for differential drift is essential. In a practical system [9.4] a resolution of 0·1°C and an accuracy of 1°C are claimed, after including the appropriate signal processing, and using signal integration times of the order of seconds.

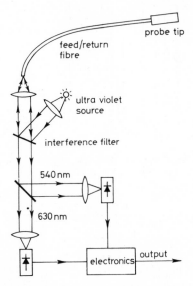

Fig. 9.4   *An optical fibre thermometer based on phosphorescence variations with temperature using interference filters to analyse the returned spectrum*

A particularly simple form of fibre optic colour probe is shown in Fig. 9.6. This uses no external sources of radiation, but simply takes black body radiation from the probe tip and transits this to the measurement photodiode. The total intensity detected by the photodiode depends only on the tip temperature, so that a very simple pyrometer may be fabricated. Detailed analysis is given in Reference 9.1. Resolution of 1° in the range 250°C to 650°C is typical. The lower end of the range is limited by detection sensitivity and the upper end by material considerations.

Another form of filter which is readily modulated by physical parameters is the Fabry Perot etalon, and, similarly, the Lyot type of polarisation-based birefringence filter. The principles of these filters are described in detail in Reference 9.5. In the Fabry Perot, the variations in the etalon separation change the transmission and reflectance functions of the filter, whereas in the Lyot filter the variation is caused by altering the birefringence of a medium between crossed polarisers as a function of environmental parameters.

The transfer function of these filters is thus a measure of the appropriate environmental parameter. The measurement problem therefore reduces to one of determining this function. There are two alternative approaches. The first is to ensure that the optical filter has the highest possible Q-factor, and to measure its physical condition by locating transmission or reflection peaks with sufficient

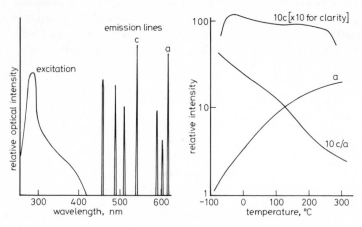

Fig. 9.5   *The phosphorescence spectrum of a rare earth phosphor, and the variation of the intensities of two of the lines with temperature*

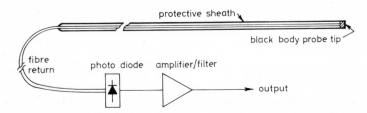

Fig. 9.6   *Simple optical fibre temperature probe, in which the photodiode itself performs the spectral analysis function*

accuracy. This implies the use of a high-resolution spectrometer, with the consequent requirement for a high degree of spatial coherence from the return fibre into the spectrometer, and a physically wide (in wavelengths) parallel beam into the wavelength analyser. An alternative, and in many ways more attractive, approach is to exploit the fact that the general form of the transmission function is well known (for instance, in the Fabry Perot it is the Airey function), and to use a relatively low-Q assembly, such that a significant amount of light is available over a wide range of wavelengths. A diode array may then sample the spectrum through the spectrometer and, provided that the system parameters outside the actual sensor remain a constant function of wavelength, this sampled spectrum may be used to compute the full transfer characteristic of the filter; hence the physical condition of the sensor may be accurately determined. The details of these calculations depend on the actual structure in use but, as an example, the reciprocal of the transmittance of the Fabry Perot filter is a sinusoidal function of wavelength (or, strictly, wave number) so that this knowledge may be programmed into the detector array computer (Fig. 9.7).

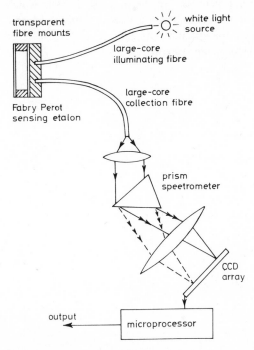

Fig. 9.7 *Spectral modulation sensor using a microprocessor preprogrammed with prior knowledge of the spectral form*

9.3 Colour probes — discussion

The colour probes discussed briefly here, and most other probes described in the literature, share a number of basic features. In all cases, ratiometric measurements are made (with the exception of the black body system), and calibration is required to establish a reference point for the ratiometric measurement. Most of the probes involve what may be viewed as 'non-standard' transmission wavelengths through the fibre. This includes both the visible and UV region and the far infra-red where the fibre attenuation is high. Probe range is therefore usually limited. The fibre also functions only as a light feed in all these devices, and again, in most cases, the probes are designed for specific applications where the use of a passive fibre feed is advantageous. There are also fibre equivalents to the Fabry Perot filter, in the form of lengths of single-mode fibre with reflecting ends or as recirculating ring interferometers. These all-fibre devices are usually configured as sensitive displacement sensors and are read using monochromatic light. They could, of course, also be read in the way described here, but at a much reduced threshold sensitivity.

The principal applications of these probes exploit small size and the passive characteristics of the technique, both electrically and chemically. Thus medical and electrical power generation are the principal areas in which these concepts — to date — have become economically attractive. Sensors based on these principles tend to be designed for specific applications, and there is little in the way of common technology between systems.

Chapter 10

Polarisation modulation in fibre sensors

10.1 Introduction

Polarisation plays an important part in a variety of optical fibre systems, especially in those in which single-mode fibre is involved. An understanding of the role of polarisation in fibre systems and the means whereby polarisation properties may be characterised is central to any discussion of fibre polarisation components and sensors. The elements of this are described in Appendix 5. The essential concepts are the use of the Poincaré sphere to trace the evolution of polarisation states through a specified birefringent medium, and the fact that any birefringent medium may be characterised in terms of two components, a linear birefringence and a circular birefringence. In fact, if we examine these two concepts closely, it very soon becomes apparent that they reduce to being equivalent. However, there are circumstances when one or the other is a more convenient way of describing a particular situation.

A variety of physical phenomena influence the state of polarisation of light. It is useful to classify these in terms of the type of polarisation modulation which may be imposed and the physical parameters which may cause this modulation. Birefringence may be introduced by a number of effects, described in the following sections.

10.1.1 Optical activity

Optical activity is an inherent property of a number of materials, and is simply a circular birefringence due to 'handedness' in molecular structure. Poincaré sphere representation of this is in terms of eigenmodes at the poles of the sphere, and the optical activity is the difference in optical phase through the material for these two eigenmodes. Thus if linear polarisation is passed through an optically active medium, the output will be rotated by an angle equal to half the phase difference between the two eigenmodes. This may also be viewed as resolving the linear input polarisation into two equal circular components of opposite signs, passing these two components as the eigenstates through the medium, and recombining them in a different relative phase.

An optically active medium is reciprocal, an important contrast to the similar phenomenon, Faraday rotation, described below. Thus if linear polarised light is passed in one direction through the medium, and is rotated by an angle θ, then light passed in the reverse direction, input at this angle θ, will emerge at the original input angle.

Optical activity occurs in crystalline materials, and in certain — usually organic — noncrystalline materials. In the latter category, the more common are the optically active sugar solutions; dextrose is so named because of its observed right-handed optical activity. Crystalline quartz is probably the best known of the optically active solids, but the phenomenon in solids is quite complex. Fused quartz is not optically active, and crystalline quartz exists in two different crystalline states (enantiomorphs) in which the structures are mirror images. One form is right handed (d-active) and the other is left handed (l-active). There are other phenomena involved. For instance, rubidium tartrate is d-active in solution and l-active when crystallised from the solution. See References 10.1 and 10.2 for more details.

Optical activity can then be used as a measure of concentrations of solutions of active materials — and has been used as such for probably over a century. In the sensing of other parameters, optical activity in crystals may be exploited. This is usually a function of temperature, and also of pressure, though the latter is of lower sensitivity. (The relative sensitivity of optical phase in fibres to pressure and temperature is the same, and the physical reasons are similar — in effect, in both cases the change in refractive index is due to changes in molecular separations.)

10.1.2 Faraday rotation

Faraday rotation is a magnetically induced optical activity. The formulation is customarily for linear input polarisation, and is expressed as a rotation of the input polarisation θ, where:

$$\theta = V \oint_0^L \vec{H}\,\vec{dl} \tag{10.1}$$

The constant V is the Verdet constant, and L is the length of the optical path in the material. By convention, V is positive when the induced activity is l-rotation when the light moves parallel to the magnetic field and is d-rotation when the light moves anti-parallel to the magnetic field. This leads to a most important distinction between Faraday rotation and optical activity. If plane polarised light is rotated through θ in one passage through the material, then returning the light in the reverse direction will result in a further θ rotation, so that on the double passage the total rotation would be 2θ, rather than zero as in the case of the optical activity.

All materials exhibit some Faraday rotation. The effect is strongest in ferromagnetic materials (and has formed the basis of, for instance, the observation of magnetic bubble domains), and is weakest in diamagnetic materials. There is a strong temperature dependence in ferromagnetic and paramagnetic materials, but a negligible temperature dependence in diamagnetics. Faraday rotation is a particularly useful effect in optical fibre sensors and component systems. It may be exploited as the basis of a current measurement probe (via magnetic fields) and also in nonreciprocal component elements such as isolators and circulators.

10.1.3 Electrogyration

Electrogyration is an electric field analogue of Faraday rotation. However, unlike Faraday rotation it only occurs in a limited range of materials, and only in the crystalline form of these materials [10.3]. The effect is sensitive to crystal orientation with respect to electric field direction, and may be regarded as electric field modulation of optical activity. The effect is present in quartz, but is relatively weak. However, it does have the potential, as yet relatively little exploited, of providing an electric field and/or voltage probe for use in the electrical power supply industry.

These are all effects which induce circular birefringence — or optical activity — under the influence of external environmental variations. There are other effects which induce linear birefringence, and these include the following.

10.1.4 Electro-optic effect

The electro-optic effect occurs when an electrical field is applied to a crystal such as lithium niobate. The refractive index of the material changes differently for different polarisation directions and propagation directions through the crystal. Depending on the crystal structure and orientation, either the longitudinal electro-optic effect (sometimes also known as the Pockels effect) or the transverse electro-optic effect may be observed. The former introduces linear birefringence as a result of applying a field along the propagation axis, and the latter as a result of applying an electric field transverse to the propagation direction. The details of the phenomenon are quite complex, and are beyond the scope of this discussion. An introductory account may be found in Reference 10.4, and more detailed descriptions in the classic text on crystals by Nye [10.5].

The electro-optic effect is an important phenomenon in optoelectronics. It is exploited particularly in modulators and switches. The modulators may be either pure phase modulators, or amplitude modulators which translate a change in polarisation state caused by birefringence modulation into an intensity modulation by passing the output beam through a polariser. Switches using electro-optic modulation are particularly important in integrated optics, where the effect may be used to fine tune the propagation coefficients in adjacent optical waveguides.

In contrast to the Kerr effect (see below) the electro-optic effect is linear with applied electric field, and thus may, in principle, be used as a means of measuring voltage. The practice is complicated by the fact that high fields are often required to produce significant effects (typically of the order of megavolts per metre) and that the electro-optic coefficients often have a high temperature coefficient, which complicates measurement analysis: Additionally, the crystal symmetry required for a linear electro-optic effect is also piezoelectric, which implies that any devices may also be pressure sensitive.

10.1.5 Kerr effect

The Kerr effect is an electrically induced birefringence which occurs in all materials. It may be thought of as a microscopic distortion of molecular symmetry, in effect lining the electron cloud slightly with the applied electric field, and thereby inducing a polarisation dependence in the refractive index. One may define two refractive indices $n_{\parallel}$ and $n_{\perp}$ parallel to and perpendicular to the applied field, and a difference in refractive index Δn given by

$$\Delta n = \lambda K E^2 \tag{10.2}$$

where K is the Kerr coefficient. In isotropic materials, K is clearly independent of direction and polarisation, but the effect also occurs in crystals, in which the value of K then becomes directionally sensitive. The Kerr effect is extremely fast, and may be thought of as instantaneous. In fact it will follow an applied optical frequency field, and as such is the most common nonlinear effect in materials. Thus, extremely fast Kerr modulator cells may be made, and discrete optic (as opposed to integrated optic) Kerr cell modulators operating at up to 10 GHz have been available for a considerable time.

Use of the Kerr effect in sensors is attractive since the effect is common and does not rely on crystal symmetry, and it is also relatively temperature independent. However, the quadratic dependence of index difference on field and the fact that the effect is relatively weak in most materials have limited its application in measuring systems.

Both the Kerr effect and the linear electro-optic effect may be characterised in terms of the K and r_{63} constants respectively. The values of these for some materials are shown in Table 10.1. The phenomena are discussed in more detail in Reference 10.4.

10.1.6 Photoelastic effect

The photoelastic effect may also be used to induce linear birefringence. Application of a stress perpendicular to the direction of propagation of a lightwave will induce an increase in the dielectric constant for light polarised along the stress direction. This effect is one which occurs in virtually all solids, regardless of crystal symmetry. In anisotropic materials the effect will clearly become directional. Stress-induced birefringence is commonly observed – and is particularly strong in some stressed plastics used for instance in stress analysis of structures – by viewing a model of the structure through crossed polarisers. Daylight is also slightly polarised, so that stress-induced bar patterns may be seen through polarising spectacles on many types of car windscreen.

In general the electro-optic coefficients are quite small, so that small stresses may only be detected by integrating the effect over a long path difference. However, the effect is commonly exploited in optics in the form of Bragg cell modulators and deflectors. The effect also forms the basis of the fibre optic hydrophone, though in the hydrophone the refractive index change is independent of polarisation until the acoustic wavelength becomes comparable with the fibre dimensions.

Table 1. *Properties of some electro-optic materials* [1]

Material	Electro-optic coefficient $(10^{-12}$ m/V)	Refractive indices[2] n_0	n_e	$n_0^2 r$ $(10^{-12}$ m/V)	Dielectric constant[3]
KDP	r_{41} 8·6	1·51	1·47	19·2	20, ∥c
(KH$_2$PO$_4$)	r_{63} 10·6			23·1	45, ⊥c
DKDP	r_{63} 23·6	1·51	1·5	53	50, ∥c
(KD$_2$PO$_4$)					
Quartz	r_{41} 0·2	1·54	1·55	2·24	4·3
	r_{63}			2·19	
LiNbO$_3$	r_{21} 3·4	2·22	2·20	16·8	50, ∥c
	r_{42} 28			137	
GaP	r_{41} 0·97	3·31		8·8	11
GaAs	r_{41} 1·6	3·34		17·7	11·5
(10.6 m)					

[1] The contents of this table should be used in conjunction with a full text on electro-optic crystal properties. The refractive index as a function of applied electric field is given by expressions of the form:

$$n_x = n_0 \left\{ 1 - \frac{n_0^2 r_{63} E_z}{2} \right\}$$

and

$$n_y = n_0 \left\{ 1 - \frac{n_0^2 r_{63} E_z}{2} \right\}$$

but due attention must be paid to crystal symmetry properties before use of the constants in the derivation of electro-optic effects for fields and/or propagation directions on other axes
[2] Refractive indices for 'ordinary' and 'extra-ordinary' rays
[3] Dielectric constants (at low – suboptical – frequencies) may be sensitive to crystal orientation; here shown parallel to and perpendicular to the crystal c-axis

The refractive index change is given by:

$$\Delta n = \frac{n^3 p}{2} \left\{ \frac{2 I_{\text{acoustic}}}{\rho V_s^3} \right\}^{1/2} \tag{10.3}$$

a relationship which is derived in detail in Reference 10.4. Here, p is the photo-elastic constant, n the refractive index, ρ the density and v the velocity of sound through the material. Clearly a figure of merit M may be defined as:

$$M = \frac{n^6 p^2}{\rho V_s^3} \tag{10.4}$$

which gives a direct measure of the efficiency of the interaction of light with a pressure wave. Values of the appropriate constants for a number of materials are given in Table 10.2.

Table 2   *Photoelastic coefficients for some materials*

Material	Density ρ tonne/m^3	Acoustic velocity V_s km/sec	n	p	M^2 normalised to water
Water	1	1·5	1·33	0·31	1·0
Fused quartz (SiO$_2$)	2·2	6·0	1·46	0·21	0·006
Lithium niobate LiNbO$_3$	4·7	7·4	2·25	0·15	0·012
Lead molybdate PbMO$_4$	6·95	3·74	2·30	0·28	0·22
Tellurium dioxade TeO$_2$: slow shear wave	6·0	0·62	2·35	0·09	5·0

[1] The photoelectric diffraction figure of merit is given

$$M = \frac{n^6 p^2}{\rho V_s^3}$$

and is here normalised to the value for water, $0·16 \times 10^{-12}$ sec^2 kg/m
[2] In many materials — especially crystals — the total photoelastic phase modulation depends on relative orientation of the input wave polarisation, the crystal axes and the applied acoustic wave. For further details see, for instance, Reference 10.4

The photoelastic effect can therefore be used as the basis for a number of transducers for monitoring pressure, strain etc., either statically or dynamically. The modulation technique may be either in terms of pure phase changes introduced by an isotropic pressure field, or in terms of induced birefringence caused by anisotropic pressures. The photoelastic effect may also induce circular birefringence when a medium is subject to torsional stress. This may be useful in, for instance, the optical fibre polarisation controller.

In terms of the Poincaré sphere representation (see Appendix 5), linear birefringence corresponds to eigenmodes diametrically opposite each other on the equator of the sphere. Thus inputting any linear polarisation, other than as one of the eigenmodes, results in general in an elliptical output polarisation (the exception is the half-wave plate, which then takes the input state to an output state also on the equator). The maximum ellipticity is clearly induced if the input state is linear at the point on the equator half-way between the two eigenstates. This would be the optimum operating point for a birefringence modulator.

We have now defined the principal means by which the polarisation of light may be influenced by external environmental variations. The distinction between modulators which cause a change in the optical activity of a material, and therefore are circularly birefringent, and those which change the linear birefringence, is a useful classification of the modulating mechanisms. Similarly, the Poincaré sphere representations of the two modulation processes serve to emphasise the differences between the two types of modulators and to simplify analysis of the phenomena. With the exception of the Faraday effect and the photoelastic effect, these phenomena occur in materials which are not readily produced in optical fibre form. In particular, cyrstalline materials are difficult to produce in fibre form, since a long thin crystal is not readily fabricated. Polarisation modulation sensors may then take the form of a fibre feed to and from a sensing element — usually a crystal — or of a continuous fibre which both transmits the light and provides the vehicle for the interaction between the polarisation state in the fibre and the external parameter to be monitored.

10.2 Polarisation modulation sensors

Perhaps one of the most successful of fibre optic sensors is the Faraday rotation current monitor [10.6], shown in schematic form in Fig. 10.1. In this system, light from a helium neon laser is launched into a single-mode optical fibre. The light is polarised before launch, and in principle maintains its polarisation direction until it reaches the current carrying busbar. Here the polarisation is rotated by the magnetic field produced by this current, and then the rotated polarisation is transmitted back to be analysed. The application of the system is on high-tension lines, where current and voltage measurements using conventional techniques (current transformers) are both expensive and difficult to implement and interpret. *re: earth loops ?*

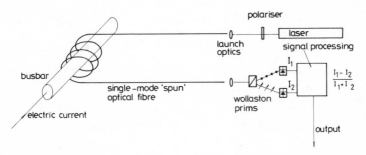

Fig. 10.1 *An optical fibre electrical current probe using Faraday rotation as the modulation process*

We have already noted that the optimum polarisation state to present to a Faraday rotation device will be linear. However, the system requirement here is that the launched linear polarisation remains so over the transmission path from the laser to the bus bar, and back again. The fibre must then be as near as possible non-birefringent over this length. In practice, all single-mode optical fibres are found to be birefringent owing to slight deviations from complete circular symmetry in the core geometry, in either refractive index or diameter or both. One could perhaps suggest heightening the tolerances in the drawing process, but a simplistic estimate of the tolerances required for less than a quarter wave of birefringence in 10 metres of fibre — which is 10^7 wavelengths — demonstrates that circularity must be maintained within this order.

This appears a formidable proposition, until it is appreciated that the principal sources of birefringence are in the deposition of the preform rod and in the drawing process, both of which can introduce slight ellipticities in the core. It is also known that these ellipticities vary fairly slowly along the length of a conventionally drawn fibre, and the polarisation characteristics of the fibre evolve in a similar relatively slow fashion (in distances of the order of tens of centimetres to metres). An ingenious solution has been obtained in the form of the 'spun' fibre [10.7]. This fibre is spun rapidly during the drawing process, so that the ellipse is rotated approximately once per centimetre. Therefore, in any given one centimetre length, there is no 'preferred axis', so that the net birefringence of this length is zero. The total birefringence of any given length of fibre is then of the order of the residual birefringence left after the 'smearing process' which takes place during spinning. This has been found to work extremely well, and residual birefringence of the spun fibre is typically less than 2·5°/metre, compared to typically several tens of degrees per metre in conventionally drawn fibre. The spun fibre has the additional advantage that the birefringence increases at a rate which is less than linear with length; thus, for the ten to twenty metre lengths required for this particular application, the total birefringence will still be more than adequate for the Faraday rotation to be accurately measured. This spun fibre is also probably the first example of a fibre specifically designed for a sensor application.

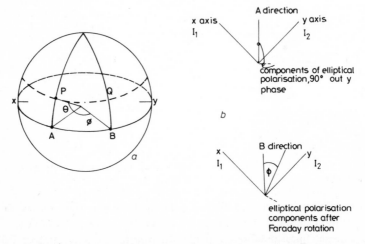

Fig. 10.2   *Poincaré sphere representation of the effect of using elliptical instead of plane polarised light as the input to the Faraday rotation sensor. The vector diagrams are used to calculate the consequent reading errors*

It is instructive to consider the effects of incorrect input polarisation on the received signal in a polarisation rotation sensor. Again, the Poincaré sphere is a convenient means of visualising the problem. Suppose that the correct input polarisation is represented by a point A (see Fig. 10.2). Faraday rotation takes this to the point B, also on the equator. This is then analysed by the Wollaston prism analyser set with its principal axes at states X and Y with A as the bisector of these two states. Suppose now that there is some birefringence in the fibre feed to the sensor. This will then put the input at point P on the sphere – here assumed to be modified by only linear birefringence. Circular birefringence in the path will simply rotate this point on the same latitude. The analyser would still be set in the same relative position – that is, on the equator at a longitude $90°$ from A. Faraday rotation then takes the output polarisation to point Q, on the same longitude as B. The ellipticity of the state of polarisation is unaltered by the Faraday effect. If the state P is aligned with the analyser as shown in Fig. 10.2b, then the outputs in intensity (remembering that the elliptical polarisation may be represented by the sum of two linearly polarised components in phase quadrature) along the 1 and 2 axes are:

$$I_1 = E^2\{1 + e^2\}$$
$$I_2 = E^2\{1 + e^2\}$$

(10.5)

where E is the amplitude on the major axis of the ellipse, ellipticity e.

If the ellipse is rotated by an angle ϕ, then the intensities are:

$$I_1 = \frac{E^2}{2}\{(\cos\phi - \sin\phi)^2 + e^2(\cos\phi + \sin\phi)^2\}$$

$$I_2 = \frac{E^2}{2}\{(\cos\phi + \sin\phi)^2 + e^2(\cos\phi - \sin\phi)^2\}$$

(10.6)

Taking the ratio of $I_1 - I_2$ to $I_1 + I_2$ gives:

$$\frac{I_1 - I_2}{I_1 + I_2} = 2\cos\phi\sin\phi\left\{\frac{1 - e^2}{1 + e^2}\right\}$$

(10.7)

In the case when $e = 0$, we have pure linear polarisation. When $e = 1$ we have circular polarisation, and the output is zero. At intermediate stages, the effect of the ellipticity is to introduce a scale factor error of $(1 - e^2)/(1 + e^2)$. Provided that this factor remains constant, then it can be calibrated out. It is readily demonstrated that the ratio e is given by $\tan\beta$, where β is the linear birefringence. Thus, for a 1% error in the perceived reading, the total linear birefringence should be less than $11°$. For 10% error in the reading, a linear birefringence of $35°$ can be tolerated. It therefore becomes apparent that the use of spun fibre over the lengths involved will introduce errors of at most a few per cent in the perceived current level. The constant error – the mean ratio of $(1 - e^2)/(1 + e^3)$ – can be corrected, but random drifts in this quantity cannot. The drift will, however, be much less than the total birefringence, so that an accurate all-fibre polarisation sensor may be realised.

The sensitivity of a Faraday rotation current sensor may be calculated from the Verdet constant of silica (3.3×10^{-4} degrees/ampere turn) and by assuming that a polarimeter with a resolution of 0.1 degrees is available. The resolution is then 300 ampere turns, so that a 10 turn coil of fibre will resolve 30 amperes. The dynamic range is determined by the detection technique. The simple ratio intensity

polarimeter is linear to 1% to a total rotation of 7°, giving a maximum current of
2·1 kA and a resolution of slightly worse than 1% over this range. With care in the
design of the analyser, the lower limit may be improved, and the upper limit may
be extended either by relaxing the linearity constraint or by incorporating some
signal processing. Practically achieved results [10.8] cover a range of currents
exceeding 10 kA with resolution of a few tens of amperes. In practice the device
has been proved stable over extended periods of operation, and has produced results
within 2% of those obtained by conventional techniques. The fibre probe has the
added advantage of high bandwidth (1 MHz) when compared with a current transformer, and this can be important in the examination of current transients under
fault conditions.

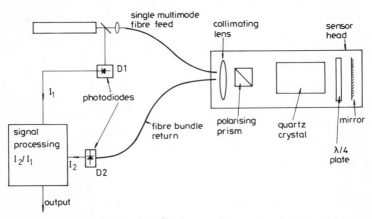

Fig. 10.3   *An optical fibre temperature probe using thermal modulation of the optical activity
of quartz as the modulation technique*

Numerous other sensors based on polarisation modulation have been evaluated,
some using free space transmission of laser light, some using fibre optic feeds
[10.9]. All except the Faraday rotation device separate the fibre feed from the
sensing function. One example is the temperature probe shown in Fig. 10.3. In this
device, light from a helium neon laser is launched into a multimode feed fibre
which is at the centre of a bundle of fibres. The light is returned to the photodetector via the remaining fibres in the bundle, and the ratio of the light returned
to the light transmitted is calculated.

The returned light is modulated by the action of the optical activity of the
quartz block, which is itself altered by variations in temperature. The laser light
has its polarisation scrambled on passage through the multimode fibre, so that the
polarisation is defined by the polarising prism. This linear polarisation is rotated
by an angle θ passing through the quartz block. This angle θ may be taken as a
function of temperature, though in general Faraday rotation is also present – a
factor to which we shall return shortly. The action of the double passage through
the quarter-wave plate is to rotate the input polarisation. If the input is aligned

at an angle ϕ to the fast axis of the half-wave plate, then the output direction is rotated by an angle 2ϕ before the second passage through the optically active material. In the second passage, the optical activity subtracts θ from the net rotation. Taking the input polarisation OZ as the reference (Fig. 10.4), the polarisation history may be traced as follows:

Output angle from first passage through quartz: $\quad \theta_f$
Output angle after double passage through $\lambda/4$: $\quad \theta_f + 2(\alpha - \theta_f)$
Output after second passage through quartz: $\quad \theta_f + 2(\alpha - \theta_f) - \theta_r$

Hence the total rotation is $2(\alpha - \theta)$. Note that, for Faraday rotation, the value of the rotation angle has the same sign in each direction and is hence cancelled.

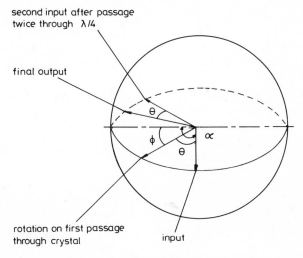

Fig. 10.4 *Poincaré sphere representation of light transmission through the sensor of Fig. 10.3.*

The polarising prism is to be used as an analyser, and so at the medium value of θ, $\bar{\theta}$, the total rotation should be $45°$ in order that the analyser operate in the linear region. Thus, the value of α should be set such that $(\alpha - \bar{\theta})$ is $22.5°$. The performance of such a device has been reported in detail [10.10] and a cylindrical probe of diameter 9 mm and length 65 mm has been found to have a resolution of $2°$C over a range $20-180°$C. Such devices are again well suited for applications in the electrical power supply industry.

10.3 Discussion

The preceding examples demonstrate some of the basic features of polarisation-based optical fibre sensors. In particular, it is usually desirable to operate the polarisation analyser in the centre of the linear region, and it is usual to assume that the input polarisation state to this device is linear. Errors — which are readily analysed — are introduced if the analyser deviates from these requirements. The system is also fundamentally sensitive to intensity noise fluctuations in the launched

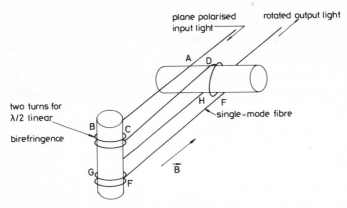

Fig. 10.5   *An optical fibre magnetic field probe which may also be utilised as an optical fibre isolator*

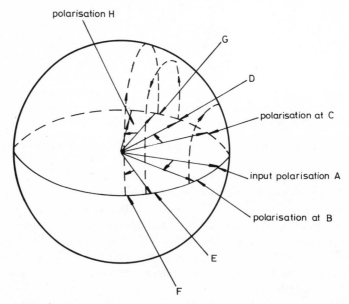

Fig. 10.6   *Poincaré sphere representation of polarisation history through the probe of Fig. 10.5. Note the alternation of the fast and slow axes prevents cumalative errors*

power and in variations in loss along the transmission path. These are compensated in most cases by taking a ratiometric measurement involving the output power from both ports of the analyser, or normalising to the launch power and assuming that no other variations occur along the transmission path. The latter is subject to variations and, for this reason, these systems may use long averaging intervals (seconds) to minimise the effect of random fluctuations.

Polarisation modulation in optical fibres may also be introduced by a number of other means. Mechanical twisting of a fibre induces optical activity, whereas wrapping the fibre around a mandrel introduces linear birefringence. These two phenomena, coupled with Faraday rotation, may be exploited to produce a fibre isolator, the principles of which are shown in Fig. 10.5. The unusual winding arrangement is used to compensate the effects of linear birefringence on the Faraday rotation. The polarisation path on the Poincaré sphere is shown in Fig. 10.6. The Faraday rotation is ineffective in the presence of significant linear birefringence.

In detection systems, it is often convenient to transfer the modulation on to some suitable form of AC carrier. In some polarisation modulation systems this may be introduced by using an electro-optic modulator to effectively change the analyser position from a positive to a negative slope, by introducing a 180° modulation between the two principal axes of the system. This is useful not only to minimise the effects of $1/f$ noise etc., but also to minimise the influence of intensity noise, which is not affected by this switching process and is therefore not transferred on to this carrier frequency.

Polarisation modulation also figures prominently in any speculation of future sensor systems. The techniques of polarisation time domain reflectometry offer potential for certain types of sensor highway, and crystal fibres could be used as multimeasurand transducers. These aspects are discussed in more detail in Chapter 13.

The principal activity in polarisation modulation sensors has been in the electrical power supply industry. The motivation here is in the availability of a transducer which does not conduct electricity, so that high-voltage insulation problems are immediately removed. It seems very likely that significant applications in these areas will soon emerge. It is also likely that applications outside these specialist industries will appear when the power and simplicity of these techniques become acknowledged.

Chapter 11

Résumé of fibre optic sensors

11.1 Introduction

It is useful to attempt to distil the information presented on fibre sensors into a form which will allow for a broad assessment of general trends and bring together some overall conclusions. These will admittedly be time dependent and will vary as the scope of available optical technology expands. Some possible scenarios for the time variations are left for the speculative chapter (Chapter 13) in which the future prospects are discussed.

It will have become evident that most measurands are compatible with optical measurement techniques. Stability, accuracy and repeatability are invariably comparable to or exceed those achieved by more conventional transduction techniques. However, it appears that until mass producible components are available, pure economics will limit the application of optical fibre transducers to situations in which they offer unique advantages over other technologies. The dominant areas of interest then involve one, or more, of high electromagnetic interference, chemically corrosive environments, hazardous atmospheres or medical applications. All these applications really exploit the fact that the transduction and transmission technologies are nonmetallic. In other applications, there are numerous measurement methods available which have considerable applications inertia. One can of course speculate that these applications may well also be eventually taken by optical techniques, though this depends on a multitude of factors which are discussed in more detail in Chapter 13.

In common with all other transduction technologies, any measurements must be made against some suitable reference, and this reference must be more precise than the required measurement. The references take a variety of forms. Differential measurements are possibly the most useful − for instance, as strain gauge bridge circuits. For highly accurate systems, provision of this reference becomes a major exercise, often involving computational facilities to compensate for the effect of changes in parameters other than the one of interest − most notably temperature − by including the effects of these parameters into a theoretical model of the measurement system.

The need for references in optical transduction systems has had repeated mention. Light sources are known to have a limited stability so that, in addition to any corrections necessary to allow for fluctuation in the modulation process with environmental variations, the actual light source must also be adequately stabilised. This may involve attention to intensity, polarisation, wavelength, spectral distribution etc., dependent upon the modulation imposed by the transducer assembly. Balanced techniques are also preferable, and numerous examples of ratiometric measurements to compensate for source drifts have already been described. Modulation techniques which cancel the noise and add the signal are also attractive for some applications. Again, in common with any other transducer technology, the source compensation question becomes more important as the desired measurement accuracy is increased.

The need for a stable reference is also implicit in the question of environmental sensitivity of the fibre to and from the sensor region. Unless the required ratio may be transmitted such that it is independent of the feed and return fibre characteristics, some considerable care is required to minimise these variations. In fact, this 'lead sensitivity' problem is apparent when it is noted that the majority of fibre sensors are either local sensing systems in which the optical signal is translated to an electrical one before transmission — for instance the fibre gyroscope and, to date, the fibre hydrophone — or are systems in which the transmission distance is limited to a relatively short range. There is considerable scope for improvement here and, again, Chapter 13 discusses some of the possibilities. Closely related with this aspect of fibre sensor design is the potential of designing passive 'smart' sensors, and some of these options are also discussed later.

The remainder of this chapter is devoted to a very brief recap of the principal features of the various modulation techniques. They are characterised in terms of the potential dynamic range, required source, detector and fibre combinations, and what may be termed as the primary measurand, along with usable secondary measurands, and a brief description of any special features of the technique.

11.2 General properties of optical modulation techniques

Intensity modulation is the simplest to implement, both in the modulator itself and the detection processes. Dynamic ranges of 50−70 dB may be anticipated, depending on the care taken in the source and detection systems. The source may be an LED or an incandescent lamp — laser sources are prone to the effects of spurious interference — and a PIN detector diode usually suffices. The primary measurand is invariably displacement, and resolution of this quantity depends on the modulation technique. Microbend transducers are the most sensitive, with resolution of fractions of an Ångström. However, these involve the use of bare fibre in the transducer, and environmental protection must be provided. Reflector motion may be detected with a resolution of the order of 1−10 nm, and mask motion across the fibre with resolution of approximately 100 nm. Displacement is the primary measurand in numerous other transduction techniques, so that current sensor technology is readily adapted to interface with intensity modulators, though provision for the intensity reference must be incorporated. Intensity modulation of light may thus be built into pressure transducers, microphones, differential pressure flowmeters, strain gauges and a wide range of associated devices. The use of coded masks, either in the form of periodic structures to increase the sensitivity and resolution, or in the form of a digital code to form an analogue to digital conversion technique, further enhances the flexibility and sensitivity of intensity modulation transducer devices.

Optical phase modulation has the advantage of remarkable sensitivity, Dynamic range of 10^7 is relatively straightforward to attain, even for static measurements. For alternating measurements using heterodyne detection, dynamic range of 10^{10} is quite feasible. There is currently some activity − again described in Chapter 13 − on extending the basic ideas to measure slowly varying parameters.

Interferometric sensors are potentially very versatile, since by using an appropriate measurand to strain convertor, a wide variety of parameters may be sensed. For instance, extremely sensitive magnetometers − using magnetostrictive coatings or magnetic glasses − have been demonstrated.

Doppler sensors are applicable only to the measurement of particle velocities. This may in turn monitor either the motion of the bodies themselves, or, indirectly, the motion of the fluid in which they are suspended. Again laser sources are required, and here the source coherence length should ideally be well matched to the distance between the system reference plane − a point which is stationary relative to the motion to be measured − and the mobile scatterers themselves. Dynamic range is difficult to assess accurately, since it depends primarily on the availability of suitable frequency to voltage conversion circuitry. This can be a very accurate process if counters are used, but simple electronics may be designed to accuracies of the order of a few per cent. PIN diode detectors and multimode optical fibres are customarily involved. Homodyne detection systems monitor velocity magnitude; the sign may only be derived using a heterodyne detection. It is difficult to incorporate the latter into the geometrical constraints determined by the required location of the reference plane. Fibre Doppler probe range is limited to a few millimetres penetration depth into the material, though this can be extended using suitably designed optics.

Wavelength distribution sensors − that is, colour modulation devices − are specially designed for a given application. The requirements are mainly that the sensing material which induces the colour modulation should offer the potential of some form of simple ratiometric measurement. There should also be a simple standard for the required measurand available to calibrate the instrument, thereby compensating effects of source spectrum drift and detector variations. Resolutions of 0·1% on the measurand may be achieved with due attention to calibration procedures, and usually the provision of a small processor with the equipment. Colour modulation is also limited in fibre range by the fact that, in most systems, the wavelengths of light fed to the sensor are well away from the fibre transmission windows. Applications of colour sensors include chemical analysis and temperature measurement.

Finally, polarisation modulation is potentially another very powerful optical fibre transduction technique. The polarisation stability of a laser source is generally desirable, though an incoherent source with a suitable polariser may be used for some applications. For devices in which the fibre itself is the sensor, monomode fibre with carefully controlled polarisation properties is required. Usually the fibre will require very low residual birefringence, though there are some sensors based upon birefringent fibres (see Chapter 13). Multimode fibre feed and return are suitable for devices in which the polarisation modulation takes place in a sensor head, and is converted into intensity modulation by an appropriate polariser/analyser combination. The resolution of polarisation sensors is component limited, and is typically in the region of 0·1% within the detection system plus whatever errors may occur in the polarisation modulator. In isotropic materials, polarisation modulation occurs through Faraday rotation, but in crystalline materials, polarisation directions may be affected by temperature, pressure, mechanical strain, and electric fields. The principal application to date has been in the monitoring of very high electric currents. There is, however, a considerable longer-term potential as more control over the fabrication of crystalline sensors is developed. There is also considerable potential in the use of polarisation modulation as the basis of transducers which monitor numerous measurands simultaneously.

11.3 Concluding comments on fibre sensors

The preceding five chapters have discussed in some detail the fundamental aspects of modulation of light in fibre sensors. These techniques are all capable of producing measurements of adequate accuracy and stability for the vast majority of applications. Some techniques, most notably phase modulation, have set new levels of detection sensitivity. Additionally the techniques also offer the required dynamic range, and in their present form are readily applied to situations involving measurements in areas where the nonmetallic probe offers an overwhelming advantage.

These modulation and detection principles are basic to most future optical fibre transducer technologies. There remain significant advances to be made at the system implementation level to fully exploit the potential of the optical medium. Many of these advances rely on technologies currently under development, though a fairly accurate picture of the future potential of optical systems may be presented. This is given in Chapter 13, and demonstates that optical transduction could play a most important role in the instrumentation systems of the future.

Chapter 12

Optical fibres in signal processing

12.1 Introduction

Optical information processing is a multidisciplinary subject which brings together a wide range of technologies and concepts. It is beyond the scope of the present text to examine the subject in detail, but it is useful to take an overall view and to identify the areas in which fibre optics and related technologies may play a part.

Until relatively recently, optical information processing was implicitly limited in scope to the use of spatial Fourier transform relationships involving bulk optics and lens systems [12.1–12.4]. Processors based on the Fourier plane concept still represent the mainstream activity in optical information processing, but the availability of guided wave optical technology both enhances the technological feasibility of a range of Fourier systems and increases the overall potential of the technology. The subject may be divided into three branches: spatial processors using analogue arithmetic, delay line processors using analogue arithmetic, and digital processors. The boundaries between these divisions are somewhat diffuse, but this provides a useful framework for the discussion in the remainder of this chapter [12.5].

The discipline exploits a wide range of other techniques outside the boundaries of conventional optics. In particular, concepts from digital systems [12.6, 12.7], especially those concerned with computer architecture and the use of higher-level arithmetic (to bases other than the two), may be directly exploited in optical computers. The discipline is evolving rapidly, and there are continuous advances made in both the technology and the concepts involved. The discussion in this chapter will concentrate on those areas in which guided wave optics may play a useful role in information processing systems. The topic of guided waves includes not only fibre optics – the main topic of this book – but also the related technology of integrated optics [12.8, 12.9]. Both technologies play a central role in optical information processing, and often the factors which determine the choice of technology are centred around size/speed considerations rather than fundamental principles. The same is also true of sensors, in that there is a significant potential role for integrated optics.

The format of this chapter is to give a brief introduction to the concepts involved in optical information processing, followed by some examples of both the techniques involved in applying guided wave optics to the concept under consideration, and of practical realisations of guided wave optical processors.

12.2 Optical information processing systems

12.2.1 Spatial processors
The principles of all spatial optical information processors are indicated in Fig. 12.1.

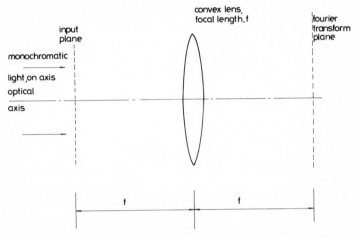

Fig. 12.1   *The basic geometry demonstrating the Fourier transform properties of a convex lens*

The input is a transparency — or optical equivalent thereof — placed perpendicularly to the optic axis of the transforming lens, one focal length in front of the lens. If the object is illuminated with a parallel beam of coherent light, aligned with the optic axis, of wavelength λ, then the optical distribution in the back focal plane is related to that in the front focal plane via the two-dimensional Fourier transform relationship [12.1]

$$F(f_x, f_y) = \int_{-\infty}^{\infty} H(x, y)e^{-j2\pi(f_x x + f_y y)} \, dx \, dy \qquad (12.1)$$

where $H(x, y)$ is a complex function describing the transmittance of the input transparency, and f_x and f_y are spatial frequencies in the x and y directions (measured in cycles per unit length). The spatial frequencies are related to the coordinates x', y' in the transform plane, and, for paraxial rays, the relationship is:

$$f_x = x'/\lambda f$$
$$f_y = y'/\lambda f \qquad (12.2)$$

where f is the focal length of the transforming lens. If the input plane and output plane are in the focal planes of the lens, then the relationship between the amplitudes is an exact Fourier transform. If the input is not on the front focal plane, there is a phase curvature in the back focal plane so that the intensity distribution in the back focal plane is the spatial frequency power spectrum of the input plane.

There is considerable literature on Fourier processors, and there are very many variations on the basic transform relationship. The variations include techniques to interface the input plane with a useful signal — usually electrical — in real time so that the speed of the Fourier processor may be fully exploited, and a great variety of architectural variations to perform other functional operations related to the Fourier transform. The most common of these are correlation and convolution processors in which the input is to be compared with some predetermined reference signal. These correlators have evolved in two basic forms with an infinity of variations — namely, the space integrating correlator and the time integrating correlator.

In the space integrating correlator, two spatial patterns are scanned across each other (see Fig. 12.2a) and the detector produces an output D which is given by:

$$D = \int_{x=0}^{L} a(x)b(x-vt)dx \qquad (12.3)$$

where v is the relative velocity of the signal and reference patterns. This may be mechanically introduced or, more commonly, the signal and reference patterns are introduced via acousto-optic modulators (Bragg cells). The zero-order stop is included in order to increase the contrast of the required signal. Note that the correlation requires time inversion of the reference, otherwise the system produces the convolution of a and b.

The time integrating correlator is shown in Fig. 12.2b. Here, the light source is modulated with one of the functions of interest $a(t)$ and the signal $b(t)$ is introduced as a spatial modulation. The spatial function is imaged on to a detector array. A given array element then detects, at any given instant:

$$a(t)\,b(x-vt) \qquad (12.4)$$

and when integrated over time, this is a correlation between a and b when the x and time axes are appropriately defined.

There are innumerable variations on these basic correlators, and these form a very important class of optical computation device. The principal difficulty with these devices has always been interfacing the data to the machine and, though much progress has been made — in the form of, for instance, light valves and tellurium dioxide Bragg modulators — the interface problem is still formidable. Successful applications of optical computation are disappointly few as a consequence, though the sideways radar tilted plane processor is an extremely elegant and practical solution to the processing of an extremely large quantity of data at high speed [12.10].

These processors are fundamentally diffraction processors, so that the use of guided wave optics is exactly contrary to the fundamental principles of the Fourier transform relationship which is at the heart of the system. It is therefore difficult to define a direct place for guided wave technology in their fabrication. However, there are numerous fibre devices which are suitable for use in image transmission applications. In particular, the fibre optic oscilloscope face-plate has been incorporated into precision oscillographs for some considerable time. The face-plate is simply an array of small fibres, typically of a less than ten microns total diameter,

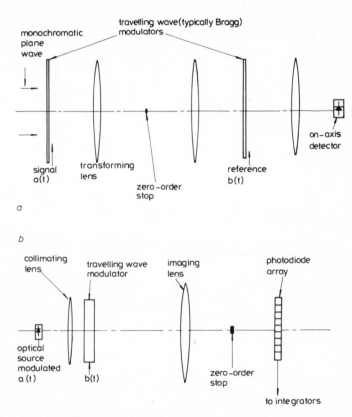

Fig. 12.2   *Optical correlators (a) space integrating and (b) time integrating*

with a core of perhaps 5–7 microns diameter, fused together in the form of a glass sheet with the cores of the fibres lined perpendicularly to the surface of the sheet. The face-plate thus transmits an image across the glass plate with a maximum point spread of the order of the fibre core diameter. This is particularly useful in precision oscillography where the image written on the phosphor by the electron beam would normally spread passing through the tube face-plate. In spatial processors, the fibre optic face-plate serves to considerably reduce the length of both stages in Fig. 12.2 – which are imaging processes [12.11]. This also eases the mechanical constraints imposed by the use of lens structures.

The dominant technical challenge with spatial optical correlators remains in the design of a suitable spatial light modulator which can be interfaced with an electrical signal. It is conceivable that some appropriate combination of integrated optics and, perhaps, acousto-optics could provide a solution which will be compact and cost effective. It is probably fair to state that previous technically satisfactory solutions have invariably proved to be extremely expensive – for instance, photoconductive devices [12.12] and liquid crystal devices [12.13].

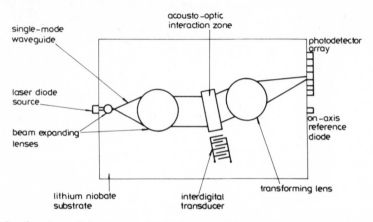

Fig. 12.3   *One form of the integrated optic Bragg modulated surface wave spectrum analyser. In other versions, the lenses may be replaced by diffraction gratings*

In fact, the beginnings of such a system can be found in the integrated optic spectrum analyser (Fig. 12.3). Here, light from a laser source is launched into a single-mode optical waveguide and beam expanded through two integrated optic lenses, either fabricated as geodesic lenses or as grating lenses, and the expanded beam is passed through an acoustic surface wave generated by an interdigital transducer (the substrate material is both piezoelectric and photoelastic – usually lithium niobate). This surface wave introduces a periodic perturbation across the guide, and thereby causes the beam in the guide to diffract. The diffraction angle is proportional to the spatial frequency of the spatial disturbance – the acoustic wave – and it is therefore proportional to the acoustic driving frequency. The intensity of the diffracted wave is proportional to the power in the acoustic beam at the corresponding frequency (at least to first order), so that the position of the output beam depends on the frequency of the acoustic wave, and the intensity of the diffracted beam depends on the intensity of the relevant component in the acoustic beam. An array of photodiodes in the focal plane of the second large integrated optic lens then acts as the input for a spectrum analyser.

There are, of course, a number of limitations to the device. The operating frequency should be in the range 500 to 1500 MHz, since at lower frequencies the wave penetration extends well below the waveguide, therefore considerably weakening the interaction. Higher frequencies start to be significantly lossy, are difficult to interface with the transducer and propagate only to a short distance within the guide, and again weaken the interaction with the optical wave. The number of resolvable frequency intervals is proportional to the number of acoustic wavelengths in the beam, and is effectively determined by the acoustic wavelength in the substrate material which corresponds to the optical beamwidth. Hence, the resolution for a 1 cm wide optical beam in lithium niobate will be approximately 500 kHz. This assumed that the lenses are diffraction limited − which they are not − and in practice, lens aberrations will cause a considerable deterioration. Typical reported performances are in the range of total bandwidth of a few hundred megahertz with resolutions of a few megahertz [12.14, 12.15, 12.16].

It is an obvious step from this device to an integrated optic time integrating correlator. However, critical analysis of the performance potential − which is beyond the scope of the present discussion − is required to evaluate the device in an intended application. In particular, linearity, dynamic range and the effects of diffraction and lens aberrations must be included.

There is, then, a significant potential for the application of guided wave optics in the spatial correlator, using both time integrating and space integrating versions. The concepts as described thus far are essentially analogue in nature, and as such are limited in accuracy. Hence an optical processor may be perhaps best viewed as a relatively coarse device acting as a prefilter for a slower, but more accurate, digital electronic machine. The basic diffraction processor concept may also be realised in a digital form [12.17].

12.2.2 Delay-line processors

Most applications of delay line processors are in radar, sonar and related systems [12.18, 12.19]. In fact, many of the processors outlined in the preceding sections are also delay line processors; there the delay was incurred through passage across the Bragg cell. In this section, the emphasis is on the use of fibre optics as the delay medium.

A figure of merit for a delay line processor is the available time delay − bandwidth product. The delay is limited at attenuation in the delaying medium, and losses almost always increase with bandwidth. For instance, in a Bragg cell operating at 100 MHz, the maximum distance which can be travelled before attenuation becomes significant may be ten centimetres (depending greatly on the choice of material). At a propagation velocity of 5×10^3 m/sec, this corresponds to a time − bandwidth product of 2000. In practice, surface acoustic wave processors may achieve a *TB* product of a few times 10^4, and bulk Bragg delay lines may be increased to the same order by appropriate choice of material and delay architecture [12.20, 12.21].

The potential of optical fibres in delay line processing becomes apparent when the possible delay − bandwidth product is evaluated. High-grade fibres will permit transmission of information over several kilometres with minimal attenuation. Additionally, the attenuation is independent of the modulation frequency on the light passing through the fibre over all realistic information rates. The fundamental limit on the time–bandwidth product is therefore set by the dispersion of the fibre. A graded index multimode fibre with a bandwidth of 1 GHz km will then have a *TB* product of 5000, but in single-mode fibres, which are material dispersion limited and are operated around the dispersion minimum (approximately 1·3 microns wavelength), the potential *TB* product is at least 500,000.

Delay lines alone are useful components, and an analysis of point to point optical fibre delay lines indicates that there are a number of applications where the small-volume, high-*TB* and EMI immunity of the delay medium are attractive [12.22]. However, the concepts only become really attractive when information can be added to, or extracted from, the delay medium along its length – in other words, if the optical fibre can be successfully tapped.

Tapping optical fibres is a feasible technology, and a variety of optical coupler techniques are available [12.23, 12.24]. In addition, integrated optics may be used as the basis of an optical coupler technology. It is inevitable that each of these tapping points introduces a loss, which will be significant compared with the overall loss of the fibre length. The loss will be due to the power actually tapped out, to connectors, splices etc. associated with the coupler, and to mode conversion which may occur within the coupler structure. The coupling coefficient of couplers for use in tapped delay line structures needs only to be very small, and if a − 40 dB coupler with zero insertion loss could be realised then even 100 taps would incur an additional loss of less than 1 dB. It is, however, much more realistic to assume a minimum insertion loss per coupler of 0·05 dB, which represents the very best achieved at microwave frequencies. In this case the coupler loss would dominate the fibre loss in the vast majority of practical cases. In effect, this imposes the limit on the available *TB* product on a tapped fibre delay line.

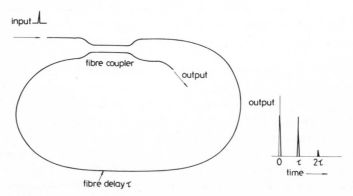

Fig. 12.4 *Recirculating delay line optical fibre filter, and its impulse response optimised for notch filter applications*

A notch filter based on fibre optic coupler technology is shown in Fig. 12.4. The impulse response of the filter may be adjusted by varying the coupling coefficient of the coupler [12.25]. For the notch filter, the coupling should be adjusted such that the first two pulses in the impulse response are equal. If K is the power coupling coefficient, then this condition may be expressed as:

$$K = (1 - K)^2 \tag{12.5}$$

which gives a required power coupling coefficient of 0·38. The third impulse then has an amplitude of 0·14, and the heights of successive pulses may also be readily calculated. The finite attenuation of the fibre is also readily included in the summation. This impulse response is calculated as the system output caused by an intensity impulse applied to the laser source. Thus, any information to be input to this filter is applied as intensity modulation. This particular format then has one immediate limitation, in that the impulse response is always positive. However, a useful filter may be realised based on these principles. The frequency response $F(\omega)$ is the Fourier transform of the impulse response and, if we neglect all terms in the impulse response apart from the first two, we find that:

$$F(\omega) = \cos\left(\frac{\omega\tau}{2}\right) \tag{12.6}$$

In a laboratory prototype of this filter, the frequency response of the filter itself showed notch depths of over 50 dB, continuing out to frequencies of over 1 GHz [12.25]. It is here that one of the advantages of fibre signal processing devices is exemplified — there is no significant deterioration in performance over the frequency ranges currently available electronically. There may be deteriorations in the signal to noise ratio, but this is concerned primarily with the optical source and detection electronics rather than the fundamentals of the filter structure. By adjusting the coupling coefficient of the coupler, this notch filter can be transformed into a bandpass filter, which in the time domain has an impulse response of — ideally — an infinite succession of equal amplitude impulses. This implies that:

$$K = (1 - K)^2 = (1 - K)^2 K = (1 - K)^2 K^2 = \dots \tag{12.7}$$

if the effects of attenuation are neglected. If we ignore the first pulse, this may be approximated by making K as close as practical to unity. Again, we here note a limitation on this basic filter format in that this particular impulse response is inefficient in optical power, since most of the available energy is in the first pulse which does not contribute to the filter response. Thus a bandpass filter would be a noisy element, and again this has been demonstrated on laboratory prototypes.

A recirculating delay line — such as that described in the previous paragraph — is limited in that each tap weight is related to the preceding ones in some well defined way, so that the impulse response is limited to a relatively few possible functions. A more common processor is the straightforward tapped delay line, in which the tap weights may be individually tailored both in coupling power and in timing. This may be implemented in fibre optic form using a sequence of couplers with variable weights spaced at appropriate intervals along the fibre. The outputs from each coupler will be summed optically on one detector.

Adjustable fibre couplers, though extremely versatile as an experimental tool, are physically large, and probably represent an intermediate step in the development of tapped delay lines. Their structure is shown schematically in Fig. 12.5. The fibres

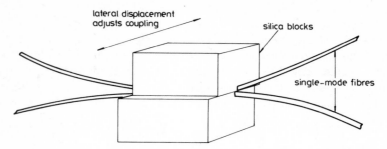

Fig. 12.5   *The optical fibre silica block coupler. The coupling coefficient is varied by adjusting the relative lateral position of the two blocks using micrometer drives*

to be coupled are cemented into a groove in a silica block on a gradual radius of curvature. The silica is then polished flat to expose the fibre cladding to within a few microns of the core. Two blocks placed together with suitable index matching oil between then allow the two fibre cores to couple. The coupling coefficient is adjusted by adjusting the relative lateral positions of the fibre cores using micrometer drives. Coupling may be varied from zero to approaching 100% in this way.

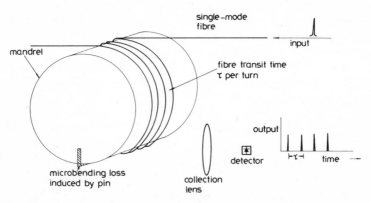

Fig. 12.6   *An optical fibre tapped delay line using microbending-induced loss as the tapping mechanism*

One alternative possibility which could achieve a tapped delay line is to use microbending induced loss as the means whereby the taps are realised. The principle of such a delay line is shown in Fig. 12.6. Light may be leaked from a single-mode fibre using a sharp radius of curvature, in the form of the tapping pin along the bottom of the fibre coil in the diagram. The tap weight may be adjusted by adjusting the tension of the fibre around the tap and by varying the position of the pin. The spatial divergence of the light emerging from the tap is minimised by the use of a high-index fluid surrounding the tapping points. The light is collected by a lens, and the tapping points are imaged on to a photodetector. This then produces an output which is proportional to the sum of the outputs from all the taps (assuming that all the taps are imaged on to points on the photodiode with equal responsivity). Further flexibility is introduced by the fact that masks may be interposed between the taps and the detector to manipulate the tap weights. An impulse response similar to that shown in Fig. 12.6 may be achieved. The power coupled at each tap is typically quite low, so that the signal to noise ratio is limited in the experimental prototypes constructed thus far [12.26].

Again, this delay line processor is based upon intensity modulation of the input light, and thus to some extent the processor speed depends on the speed with which the source may be modulated, which may be up to 1 GHz with readily available technology. The time delay between taps may be as short as a few centimetres along the fibre, corresponding to hundreds of picoseconds of delay. Therefore, efficient filters for operation well into the microwave frequency band are feasible with this technique, provided that the laser modulation frequency is compatible. (Very fast photodetectors are available [12.27].) One other possibility for this class of processor is to use pulse modulation of a laser source. Lasers with pulse durations of tens of picoseconds are commercially available, and there is a rapidly expanding interest in femtosecond optics involving light pulses of a few optical wavelengths in duration [12.28]. This leads naturally to the use of these delay line filters as digital rather than analogue processors, and we shall return to this topic later.

Intensity modulation processors, regardless of the format, are limited by the fact that intensity is, by definition, a positive quantity, so that the impulse response is always positive. This restricts the range of applications, both analogue and digital, to transfer functions which are even in frequency space. There are, of course, many other properties of light which may be modulated, and in essence one is seeking a property which may be of either sign or preferably complex so that a full range of transfer functions becomes available. Polarisation and phase modulation are the principal alternative contenders.

One could argue that orthogonal polarisation states may be used to represent positive and negative values, and certainly this is compatible with the design of polarisation-sensitive taps to represent positive and negative quantities. Perhaps polarisation retaining fibre could be used as the transmission medium? However, the detection scheme — which ideally is performed optically — should be such that the addition of positive and negative quantities accounts for signs. There is no way in which two orthogonal polarisation states may perform this operation without phase coherence. It is, of course, feasible to direct the polarisation states on to different detectors and perform the signed addition electronically.

We are, therefore, led to the inevitable conclusion that the only modulation technique which permits the transmission of sign and amplitude information is phase modulation. The use of phase modulation implies that any detection must be performed interferometrically, and that the optical source to be used is coherent over the time intervals of interest.

A number of interesting guided wave signal processors based on phase modulation of coherent light have been built, and a considerable number more have been proposed [12.29, 12.30]. These have been invariably based on integrated optics rather than fibre optics as the guiding medium. Phase modulation is particularly compatible with integrated optics, since a common integrated optical substrate medium – lithium niobate – is electro-optic; optical phase may thus be modulated by placing electrodes along an integrated optic guide. The phase delay thereby induced may be written as:

$$\phi_d = Vl \times \text{constant} \tag{12.8}$$

where V is the applied voltage, l is the length over which it interacts with the guide, and the constant depends on the geometry of the guides and electrodes and the electro-optic coefficients of the substrate material.

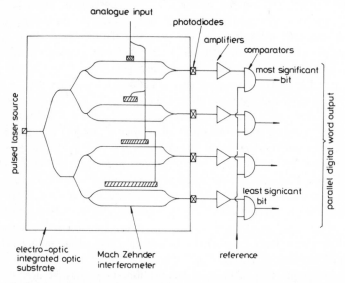

Fig. 12.7   *An integrated optic analogue to digital convertor based on electro-optic modulation of a bank of Mach Zehnder interferometers by the application of the signal to be digitised to the interferometers*

This forms the basis of a wide variety of signal processing devices. An integrated optics analogue to digital convertor is shown in Fig. 12.7. This has been demonstrated as a four-bit 828 megasamples/sec device [12.31]. The principle is simply that each of the interferometers produces an output given by:

$$I \propto (1 + \cos \phi_d) \tag{12.9}$$

The lengths of the electrodes on each successive interferometer are related by factors of two (the shortest two may be of equal length and may have a static difference of $\pi/2$ applied between them in addition to the electrical signal to produce a Gray code). The output from each of these interferometers is applied to a comparator with a threshold of half the maximum power point, and the interferometer outputs are related to the applied voltages as shown in the figure. Examination of the outputs show that the result is a four-bit representation of the input voltage.

Alternative coding formats are possible by varying the electrode formats and the DC bias voltages applied to each interferometer. An integrated optic analogue to digital convertor offers numerous advantages including very high spped, complete isolation between the input and outputs of the system, and compact format.

Extremely fast sampling systems are also feasible in integrated optics. Pulses with total widths of the order of picoseconds may be generated using numerous techniques. A microwave frequency directional coupler modulator with a microwave frequency bias applied to it may be arranged to be in synchronism for only a small fraction of the microwave cycle [12.32] (Fig. 12.8). The relative dispersion characteristics of the two guides must be correctly designed so that the switching takes place when the applied field passes through zero, and therefore has the maximum time derivative. In principle, subpicosecond sampling times are possible with this device.

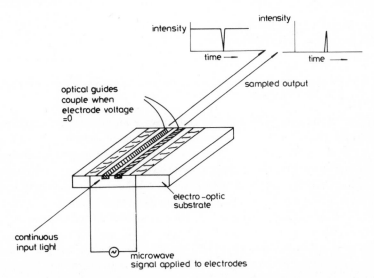

Fig. 12.8   *An electro-optic short pulse modulator. Power is only coupled to the output guide when the applied voltages to each guide are equal. If the voltages are a microwave signal applied in antiphase, then the coupling point travels along — assuming the microwave and light signals travel in synchronism — producing a very short output pulse whilst the microwave voltage crosses zero*

An alternative scheme [12.33] relies on transmission of the optical signal through successive interferometers (Fig. 12.9), each phase modulated with a harmonically locked alternating signal with a factor of two frequency difference between the various inputs. The output is then clearly:

$$\frac{P_{out}}{P_{in}} = \Pi^N \cos^2 \omega_n t \qquad (12.10)$$

when the biasing amplitudes at the frequencies ω_n are chosen such that each phase modulator imposes a peak to peak phase shift of π. (The same effect can be obtained if the frequency applied to each phase shifter is the same, but the amplitudes are increased by a factor of two from one shifter to the next). Again, this device is capable of operation at gigahertz sampling rates, and low-frequency devices operating on this principle have been demonstrated.

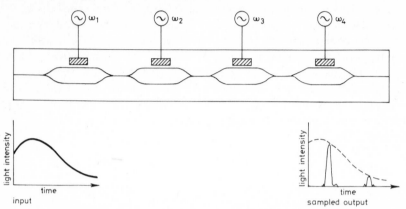

Fig. 12.9 *A high-speed optical sampler using cascaded interferometers to periodically chop an incoming lightwave*

Another interesting application of guided wave interferometers is as frequency shifters/modulators for use in heterodyne interferometer schemes. The simplest of these is the serrodyne [12.34] modulator, which applies a linearly advancing phase modulation to a wave propagating along a single guide. The concept may be modified to minimise spurious sideband generation [12.35], and detailed analysis of the device may be found in these references and in Reference 12.36.

There are numerous other single-sideband modulation techniques available in integrated optics form. An optical analogue of the radio frequency quadrature modulator is a viable alternative [12.37]. This operates by using a square-wave modulation in phase of amplitude 90° (which is exactly identical to a square-wave unit amplitude modulation) applied in quadrature to two arms of an integrated optical interferometer (Fig. 12.10). The output spectrum contains the required sideband and components which are spaced by four times the modulation frequency; therefore, minimising the detected outputs on the monitoring port at ω_m and $2\omega_m$ effectively ensures that the bias point of the modulator is optimised. Yet another integrated optic frequency shifter uses the modulation of coupling between dispersive guides as the frequency shifting technique [12.38].

This has given some indication of the potential of integrated optical signal processing, and of some of the many functions which may be implemented using these techniques. Of course, this is a slight deviation from the fibre optics systems which are the subject of this book, but the principles are very similar, and it is to signal processing with phase modulated light in fibres that we now turn.

The attraction of phase modulating light within an optical fibre is that the modulation process may be effected completely without optical loss. Numerous variations on the optical fibre phase modulator have been described, but all simply involve the attachment of a fibre optic waveguide to a piezoelectric material, which may be crystalline or a suitable polymer film. In common with the previous signal processing concepts, a single-mode fibre is usually required. The piezoelectric crystal is electrically driven — usually at its mechanical resonant frequency — and as it vibrates it also stretches and compresses the fibre attached to its surface. In the most convenient form the modulator will be a cylindrical crystal with several turns of fibre attached to increase the modulation sensitivity. Such modulators are capable of imposing several thousand radians of phase deviation at the mechanical resonance. However, the frequency range is limited to less than a few hundred kilohertz.

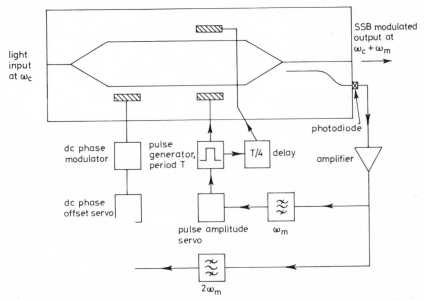

Fig. 12.10:   *An integrated optic single-sideband modulator incorporating feedback networks to optimise the output spectrum*

At higher frequencies, plane-wave transducers may be used to introduce phase modulations, at frequencies up to 100 MHz. Above this value, there are numerous complications owing to the inevitable losses of the various interfaces between the crystal and the fibre core, though with careful design it is in principle possible to impose phase modulations at frequencies up to the gigahertz range [12.39, 12.40].

The phase modulation effect may be exploited as a transversal filter in the form shown in Fig. 12.11. The taps are spaced at the required intervals and clipped on to the single-mode fibre line. The weights in both phase (delay) and amplitude may be readily adjusted by adjusting the acoustic drive to each of the phase modulators. The phase delays represent a fine tuning of the line delays, and also offer the possibility of dispersive processing which is required in many applications. Therefore the phase modulated highway offers more flexibility than the optical tap but, since it does rely on ultrasonic modulation, is probably limited in signal frequency compared with the previous examples.

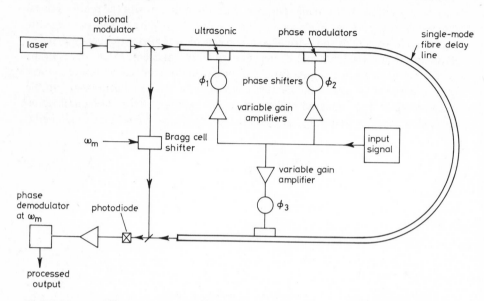

Fig. 12.11   *A coherent single-mode optical fibre delay line processor with the capability of complex taps based on ultrasonic phase modulation of the light in the fibre*

The filter system shown in Fig. 12.11 uses a Bragg frequency modulator as the the source of a local oscillator signal, which has to operate at a higher frequency than any used in the processor. This shifts all the phase modulation on to an intermediate frequency carrier and thereby eliminates the problems due to slow drifts in relative phase of the two arms of the interferometer. A phase demodulator optimised for use in the processing frequency band will then see the system as a filter with an impulse response determined by the spacing of the taps and the electronically controlled tapping weights. Further functions may be incorporated into this processor by including a phase modulator at the light input. Operations such as correlation and convolution between the input and the modulating function are then realisable.

A somewhat different delay line processor, capable of performing matrix multiplications, is the systolic array shown in Fig. 12.12 [12.41]. In this case the optic couplers are arranged to have coupling coefficients k much less than unity, and the input data is in the form of intensity modulated light pulses into a single-mode optical fibre. The delays between the couplers are arranged to be identical, and the input pulses are spaced by this delay time. The output at the detector ports is readily shown to be a train of pulses of intensity:

$$a_1 k_1 , a_1 k_2 + a_1 k_1 , a_1 k_3 + a_2 k_2 , a_2 k_3 \qquad (12.11)$$

The first and last of this sequence can be ignored, but the middle ones repesent the operation:

$$\begin{bmatrix} b_1 \\ b_2 \end{bmatrix} = \begin{bmatrix} k_2 & k_1 \\ k_3 & k_2 \end{bmatrix} \begin{bmatrix} a_1 \\ a_2 \end{bmatrix} \tag{12.12}$$

The concept may be readily extended to multiply any column matrix by a square matrix that has identical diagonal elements.

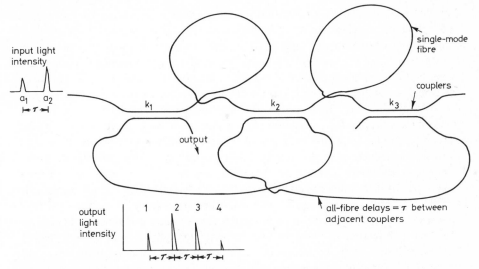

Fig. 12.12 *An optical fibre systolic array analogue multiplier matrix multiplication of for real positive numbers*

This particular signal processor is then capable of performing numerous important operations, most notably the discrete Fourier transform, in which the a matrix is the sampled data and the b matrix the Fourier components [12.42]. However, the Fourier transform operation implies the use of complex number operations, and in its simplest form the systolic array processor produces only real multiplications. Again, in principle, complex operations may be realised by using phase to code the argument of a complex number. This phase may be either optical phase or the phase of some carrier frequency — the latter having the advantage that a high degree of optical coherence is not required.

In common with most other optical processors, this array has the advantage that it can operate on data in real time, with the input pulse train derived by sampling an analogue signal. The delay required is defined by the necessary sampling rate on the input signal, and the length of the array by the required resolution. There are, however, two aspects of the system which require careful analysis before such a device can be applied. The first is that fact that both the values of k and the delays between the various elements will involve errors, and in particular the multiplication constants are subject to being defined by to, at best, 1% accuracy. There is also the fact that the simplified account of the multiplication process ignores the effect of the power coupled out at each tapping point. This is only acceptable if this effect is less than the tolerance of the operation. In practice, both these effects may be readily analysed, and the characteristics of the multiplier may be assessed. The conclusion is that the concept is limited in accuracy and dynamic range in a similar way to most other analogue optical processors. Again, the application may be viewed as that of a high-speed preprocessor with later electronic digital analysis of data of interest.

This then has described the use of fibre optics as delay line media. The advantages of a fibre optic delay line lie in its high time—bandwidth product (especially for single-mode fibres), in its compact form, and in its immunity to electromagnetic interference. In common with any delay line medium, the power of the delay may only be fully exploited if suitable tapping technology is available, and in particular if complex number representations are feasible. The only currently available technology which is readily tapped is the ultrasonic phase modulated device. The remainder have, in effect, mechanical taps which are difficult to reproduce to a high accuracy and are implicitly only capable of real positive operations. However, the former is relatively slow, limited by the ultrasonic frequency, whereas the latter may be considerably faster. Delay line processors are at a very early stage in their development, and much research remains to be completed before a full assessment of their potential may be made. The delay line processor also involves a critical assessment of the required source properties for optimum performance. In particular, the architecture of these devices involves the splitting and recombining of signals with time delay. In these circumstances, the phase noise properties of optical sources are especially important, even though the systems are themselves based upon incoherent concepts.

12.2.3 Digital processors

The central element in an electronic digital processing system is a nonlinear element, usually a bistable, and numerous of these elements are interconnected to form the required logic function. The bistable is immediately compatible with binary arithmetic, and also implies a useful degree of noise immunity in the electronic system. Even so, in electronic processors there is now considerable interest in extending the arithmetic operations to base four or higher since, in many situations, the available signal level is effectively 'wasted' when used to perform binary arithmetic [12.43]. In optical processors the noise immunity is significantly higher than in electronic processors, and so in the longer term there is a compelling interest in evaluating digital systems with multivalued logic elements [12.44].

This, of course, presupposes the existence of suitable multilevel switchable optical memory elements. Considerable effort has been expended on the design of such elements, though to date a fully practical solution remains to be realised. There are two possible approaches to the design of an optical memory element. The first is to utilise optical nonlinearities within a material, and the second is to use an electro-optic combination with electronic feedback to induce effective nonlinearities.

A Fabry Perot interferometer may be used as the basis of both types of device. The interferometer consists of two parallel reflecting surfaces separated by some convenient distance, typically in the region of a centimetre [12.45]. If mono-chromatic parallel light is used to illuminate the interferometer, then the output from the interferometer as a function of plate separation will follow the form shown in Fig. 12.13. Suppose that this interferometer is tuned such that the output is at a transmission peak. The total round trip path difference through the cavity is then of the order of several thousand wavelengths in a 1 cm cavity. If the path difference is changed by an amount $\gamma/2$ (see Fig. 12.13) then the output from the interferometer will be halved. Typically, γ corresponds to a shift of approximately 0·1 optical wavelengths, so that a change in the cavity length of one part in 20,000 results in switching the cavity from fully 'on' to 50% 'on'. In principle, this change in length could represent binary states.

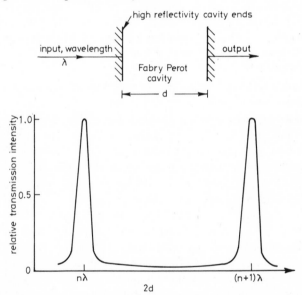

Fig. 12.13   *The Fabry Perot interferometer and a typical variation of transmission through the interferometer with optical path between the two mirrors*

This very simple estimate leads to the conclusion that a change in dielectric constant of the material within the cavity of the order of one part in 10^4 is adequate to produce a bistable optical device. The decision is then how to implement this change. Many materials exhibit optical nonlinearities, so that a cavity fabricated from a nonlinear material could form the basis of an optical switch. In most cases the nonlinearity will exhibit some hysteresis, so that a memory element is in principle feasible. Numerous experimental observations of these effects have been reported [12.46] and the field is one which is rapidly developing. However, there remains one important difficulty. The required induced nonlinearities imply power levels in a single-mode optical cavity of the order of 10–100 mW, and this is excessive as a 'per gate' requirement. There is some interest in the use of organic materials [12.47] in which the nonlinear coefficients may be up to fifty times higher. However, though the 'per gate' optical power is significantly lowered, it is debatable whether even this improvement is adequate for a complex computer system.

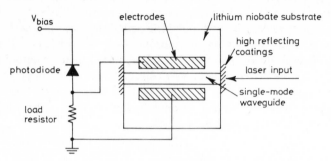

Fig. 12.14   *A Fabry Perot bistable incorporating feedback from the detected output to the optical path within the cavity. The speed of the device is limited by the effects of stray capacitance in the circuit*

The alternative technique is to use an electro-optic feedback system. This is shown diagrammatically in Fig. 12.14. The phase modulation is now effected electronically via the photodetector feedback loop. Thus, if we consider the condition where the cavity is in an 'off' position with no bias, then increasing the optical power into the cavity turns the detector on and thereby adjusts the effective cavity length, until, at a critical power input, the phase bias is exactly correct for the device to switch fully 'on'. Clearly the feedback loop should be appropriately adjusted, but a bistable element may be realised in this way [12.48]. Many more complex variations of the same concept have been reported [12.49].

These are but two of many possibilities for the creation of an optical bistable element. It seems likely that the electro-optic hybrid will find uses in the near future, though many of the attractive features of an all-optical-domain computer are undiminished by the high power requirements of these all-optical devices. The technology of guided wave optical logic is still at a very early stage in its development.

It is probably also appropriate to mention that a multichannel Fabry Perot interferometer can form the basis of a useful spatial digital processor [12.50]. A spatial phase object placed within the plates may be effectively digitised by the action of the interferometer. An image of the object illuminated in this way will contain either 'on' or 'off' portions depending on the phase of the object at the point of interest. The spacing of the plates and their reflectivity determines the resolution of the system and the significance of the bit in the binary output.

Any interferometer may, of course, also be considered as the basis of a logic gate. The Fabry Perot has the advantage of an abrupt transition between 'off' and 'on' states, whereas most other interferometers exhibit a cosinusoidal form of the dependence of detected output on interferometer path difference. In fact, the analogue to digital convertor mentioned previously is a good example of such a device. Using different electrode configurations on the same basic format will synthesise a variety of logic fluctions.

The use of digital concepts in optical signal processing requires not only the development of effective gate structure but also investigations into computer architectures for use in such systems. Binary gates are feasible, but multilevel gates would be especially attractive. However, this then enters the realm of the design of switching elements suitable for use with multilevel arithmetic. The potential of a compact high-speed, low-power computer operating at room temperature remains, but the necessary fundamental research to evaluate the concepts has barely commenced.

12.3 Discussion

The objective of this chapter has been primarily to indicate the considerable potential of guided wave optics in signal processing. The use of optical techniques in signal processing has been discussed for over quarter of a century but, as yet, applications have been relatively few and highly specialised. The principal problems with these traditional spatial optical processors lie in the input and output of data from the system in real time, and in the mechanical tolerance required on the optics.

Changing the emphasis to guided wave optics improves the overall applications potential considerably. It is easier to modulate a guided wave electronically, and mechanical requirements are relaxed. In principle, then, the possible role of optical signal processing should be widened considerably. For spatial processors, the place of guided wave optics appears to be in the use of coherent fibre bundles to transmit and/or manipulate spatial data to improve the spatial precision of certain input/output devices. Delay line processors are an immediately obvious application of optical waveguides, though the concepts require the availability of a complex tap with a high bandwidth before the immense gain—bandwidth product potential of the delay medium may be fully realised. On the digital level, advances are required on both gate structures and computer architectures suitable for a fully optical digital computer.

The previous paragraph has obviously somewhat oversimplified the situation in order to identify important areas. Perhaps it is also important to recognise the fundamental limitations of guided wave optics in signal processing. The final limit is, of course, shot noise, but it has already been demonstrated in the discussions of sensors that this is rarely the actual limit. In signal processing the technological limits remain to be characterised, so that a realistic assessment is extremely difficult. The principal questions concern the design of taps to single-mode optical fibres, and both the technology of logic gates and the architecture of digital optical computers. There is a little more speculation in the concluding chapter of the book, since not only is the optical computation realisable but also could be integrated with an optical sensor, so that the technologies may closely interact. But exactly how – this remains to be seen.

Chapter 13

Some speculations

13.1 Introductory discussion

The preceding chapters have demonstrated that most transduction processes and many computational processes may be implemented in the optical domain. The use of optics implies high speed, immunity to electromagnetic interference, often long-distance transmission potential, small physical size and numerous other advantages. The question as to where and how these techniques may be useful has been left — perhaps appropriately — to this chapter. An objective assessment of the real potential of these technologies must start with the assertion that most of the functions which can be realised optically may also be realised using more well established (and therefore, in most eyes, preferable) technologies. The objective of this final chapter is then to describe some applications in which optical systems have real and identifiable advantages over any competing technologies, and also to identify some important technologoical and conceptual advances which will take place in the near future and examine how these may influence the future of optical sensing and signal processing.

There are four principal aspects to this discussion. The first is to examine technological potential, with the constraint that the market for any systems which are within the remit of this book will be relatively limited, so that only relatively minor technological developments can be considered specifically for these applications. However, the background of activity, in the telecommunications arena in particular, is most important. This will without doubt expand the range of available components from which sensor and signal processing systems may be assembled. One then has to turn to the possible exploitations of these new tools. It is feasible to classify the potential into the subjects of sensors and signal processors but, as will become apparent, a third category of hybrid systems involving combinations of any two, or all, of sensing, signal processing and data transmission should be included. Not only that, but the boundaries become blurred. However, we can most certainly establish that there is a considerable scope for future applications of optical techniques.

It is also useful, before entering into this discourse, to review the properties of the available medium, namely the modulation of light, and the transmission of modulated light along optical fibres. This is particularly interesting in that it reveals most forcibly that current sensing technologies do not even remotely approach the full exploitation of the medium. Even the simplest of optical communications systems, based on a low-power light-emitting diode, step index fibre and a simple PIN diode receiver, is capable of transmitting a bandwidth of the order of megahertz with perhaps 40 dB signal to noise ratio over a transmission distance of several hundred metres. State of the art systems are now transmitting hundreds of megabits per second over distances exceeding a hundred kilometres. Clearly the bandwidth and transmission distance requirements of most sensor systems are negligible compared to the performances achieved in PTT applications. One is therefore led to the inevitable conclusion that the transmission medium is overprovided for the majority of these applications. It is interesting to compare this with the majority of other industrial instrumentation systems in which the bandwidth/distance capabilities of the various components are usually closely matched.

There are some optical fibre sensor systems — most notably the gyroscope, and to a lesser extent the phase modulated hydrophone and magnetometer — for which the transmission aspect of sensing with optics is of relatively minor importance. These devices stand on other merits of fibre sensing — for instance, interference immunity, high sensitivity or geometrical flexibility.

In signal processing applications the emphasis is somewhat different. The fundamental motivation for the use of fibres in signal processing is the high potential time–bandwidth product, which is exactly equivalent to the high-bandwidth communications aspects mentioned earlier. Exploitation of the concept is currently restricted by the available technology to interface the fibre with a high-speed data input and to suitably manipulate the data within the fibre.

Optical instrumentation technologies involve another crucial aspect, which has thus far been ignored. It is, at least in principle, possible to use optics in a multi-dimensional fashion to perform several functions simultaneously. Any discussion of the future role of optics in instrumentation and signal processing should endeavour to incorporate all these aspects of the potential uses of light.

13.2 Techniques and technologies

There is currently a considerable background activity in the general areas of optical systems and fibre optic components, spurred primarily by the immense potential for such systems in PTT activities. Many of these activities are immediately applicable to sensing and signal processing, either as direct improvements to existing concepts or to enhance the range of optical techniques into regions inaccessible to other competing technologies.

The principal current research interests in PTT optical systems lie in the use of the so-called longer-wavelength systems in monomode fibre where dispersion becomes minimal and attenuation is also extremely low. This implies that inevitably sources, monomode fibres, detectors and connectors will all become available for use in this band [13.1]. However, PTT systems are invariably point to point links, so that splitters and combiners are not so immediately important. The other aspect of these systems is that their cost needs only to be compatible with the application – namely the long-distance transmission of vast quantities of data – and this may not meet the economic constraints of other systems unless the full fibre potential is actually used.

Another important area of activity in fibre optics is in the development of polarisation retaining fibres which use either elliptical cores [13.2] or a mechanical stress profile [13.3] to induce linear birefringence. The fibre thus has two principal axes, and linearly polarised light launched into the fibre along one of these axes will emerge from the fibre in the same polarisation, provided that the coupling between this axis and the orthogonal one remains negligible. Light launched into both principal axes simultaneously can emerge in any arbitrary state of polarisation, depending on the differential phase path between the two axes at the end of the fibre.

Integrated optics is another technology of central importance to signal processing and sensing applications. It is also of some relevance to the PTT industry, for devices such as wavelength multiplexers, splitters and combiners, and switches [13.4]. It also seems probable that the PTT industry will solve one of the important problems with integrated optics, which is that of interfacing with a fibre optic guide. The symmetries of integrated optic waveguides and fibre waveguides are clearly totally different. However, the problem is somewhat allieviated at longer wavelengths, and acceptable loss figures (better than 1 dB) are beginning to be regularly reported [13.5]. Integrated optic devices, both active, based on lithium niobate substrate material, and passive, based on glass, are potentially extremely useful in instrumentation, especially if the light source can be supplied and the light modulation detected remotely via a monomode fibre.

Light sources, especially semiconductor lasers, are currently developing at a considerable rate. Emphasis on coherent systems using frequency or phase modulation of light is increasing – in fact it can be argued that all optical communications systems developed to date use the optical equivalent of a spark gap radio transmitter as the source. Coherent systems require stabilised sources and a stable local oscillator signal at the receiver. Preliminary results with laser stabilisation appear to be promising, and this is one area in which significant developments are to be anticipated in the next few years [13.6]. The requirements here are for the suppression of laser noise, especially phase noise, the stabilisation of the laser oscillation frequency into a genuine single longitudinal mode, and a suitable means for a controlled frequency or phase modulation of the source. Such techniques will become available, and their implications for use in sensing and signal processing systems are significant.

Yet again, the longer wavelengths will become important, and it may become imperative to use this part of the spectrum, simply for the available component set, despite difficulties in viewing at these wavelengths and in performing low-noise electronic detection. Again, though, the PTT industry is likely to provide solutions to these problems.

Detection techniques may well also change in the foreseeable future. In particular, the use of heterodyne techniques with a real local oscillator (instead of the more usual local oscillator derived from the optical source) could considerably change the emphasis on the design and use of photodetector circuits. Shot-noise-limited detection could therefore become the norm, and a highly sensitive receiver module may become a readily available item — analogous to the heterodyne receivers in radio frequency circles.

The optical component set is then expanding at a considerable rate, and it is probable that the available tools for a system designer will increase significantly in the next decade. There are also developments on other fronts which could prove to be similarly important. Primarily, these concern techniques for micromachining. In two dimensions, standard photolithographic processes continue to improve in resolution and repeatability. The scope for the use of these techniques in optical systems is considerable, and is as yet barely exploited. Perhaps yet more intriguing is the development of three-dimensional machining techniques based on anisotropic etches in silicon [13.7]. Silicon may be formed into intricate and well controlled shapes using this technique. Silicon has also recently entered the picture as a potential material for optical integrated circuits. Silicon is relatively transparent at wavelengths longer than 1·1 microns, and has another interesting property in that it is relatively nondispersive from very low electrical frequencies (effectively DC) to wavelengths corresponding to photon energies just below the bandgap. Therefore there exists the possibility for the use of travelling wave interactions between a low frequency and an optical wave within the material, without the use of complicated phase matching systems. This appears to have great potential.

Doubtless other technologies will also prove to be important. There is clearly scope for the use of novel materials in optical systems to convert external stimuli into optical modulation. Holography is another optical tool with considerable potential in these applications. In particular when, or if, the adaptive hologram responding to low light levels becomes available, this could have a considerable impact on both sensing and signal processing.

The remainder of this chapter is devoted to mapping out a number of developments which are technically feasible — though perhaps technologically challenging — and which could significantly enhance the role of optical techniques in instrumentation. Many of these systems to some extent mimic presently available systems based on alternative technology, but with the speed, transmission distances, immunity to electrical interference and zero electrical power requirements which can only be afforded by using light as the communications medium. Yet other systems open the way for new concepts in instrumentation. The objective in the discussion is primarily to indicate that the exploitation of light as a medium in these applications has barely commenced. However, the potential can be tapped, and will result in elegant and useful systems which will offer facilities which cannot be obtained in other technologies. This should broaden the range of applications of optical techniques in sensing and signal processing by significantly enhancing the economic performance value. So, now let us indulge in some speculations.

13.3 Optical fibre sensor systems

It will have become evident in the first chapters of this book that optical techniques, coupled with fibre optic feeds or fibre optic feeds or fibre sensing, are capable of performing the majority of measurements required in industrial and military environments. The techniques offer real economic advantages in only a relatively small number of applications, including regions of high electromagnetic interference (for instance the electrical power supply industry), medical monitoring, where an inert sensor is especially useful, and remote and/or hazardous areas, where low attenuation and low or zero electrical power consumption is attractive. According to recent estimates, these applications could occupy 2% of the market by 1990 [13.8]. In more conventional environments, there is little or no advantage to using a present generation optical sensor.

It is interesting to examine what may be the fundamental reason for this. Optical techniques are evolving primarily for long-distance high-volume data transmission. Consequently, many of the readily available components are capable of high-volume data transmission. In these applications the price/performance ratio is excellent but, in the low-data-rate applications typical of process instrumentation, the use of fibre as the transmission medium to a single sensor element is a complete underutilisation of the medium. To illustrate this, applying the techniques of Chapter 6 we may calculate that the signal to noise ratio at a PIN diode receiver in a 1 kHz bandwidth for a receive optical power of 10 microwatts is greater than 90 dB. To read a sensor with 0·1% resolution requires a receiver signal to noise ratio of about 40 dB, depending on the assumptions made about the receiver (in principle only 30 dB would be required [13.9]). There is thus 50 dB of 'spare' signal to noise ratio available in a very simple optical system. It can be argued strongly that unless at least some of this extra signal to noise ratio is put to good use, the applications of optical measurement systems will continue to capture only the 2% of the market which involves specialist applications. Exploiting this additional channel capacity increases significantly the usefulness of the feed and return channel, which is, in many applications, the principal expenditure in the system.

It should be emphased that here we are addressing industrial applications, and Fig. 13.1 shows a simplified schematic diagram of a general transducer as used in these applications [13.10]. This figure also, of course, emphasises the relatively small difference between the 'conventional' transducer and an optical one. We have simply replaced straightforward copper feed and return leads with optical fibres (which are usually more expensive, especially for very low bandwidths). The discussion changes in emphasis when one comes to consider 'specialist' fibre sensors – for instance, the gyroscope and the hydrophone – of which more later.

There are a number of obvious means by which the 'surplus' signal to noise ratio could be put to good use. The most obvious one is some form of transducer multiplexer [13.11] which several sensors feed on to one optical communications channel. Other possibilities include the use of remote passive analogue to digital convertors, possibly with other passive signal processing incorporated at the transducer, and if appropriate, the availability of high-resolution remote measurements. These are effectively hybrid systems and are discussed in more detail in Section 13.5.

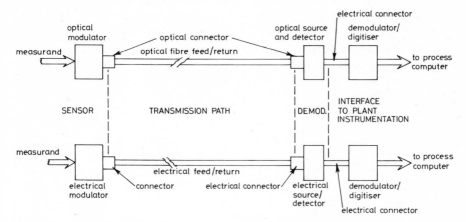

Fig. 13.1 *Comparison of an optical fibre transducer which may be applied to process instrumentation, and an electrical transducer performing the same task*

Transducer multiplexing may be divided into two broad areas: that of multimeasurand transducers, and that of true multiplexing systems where several devices are fed on to one data link. The former may be defined as those devices in which different properties of light are modulated by measurands that are different but may be all at single location. The latter includes systems where the same optical property is modulated, but at different physical locations, and each location is separately read by a suitably designed optical system.

One particularly elegant form of multimeasurand transducer system [13.12] is shown in Fig. 13.2. In this device, an electro-optic fibre is proposed for use as the transmission path between ground and a high-voltage transmission line. The extent of the linear birefringence of this section of fibre is proportional to the total voltage along the fibre path. Thus, if the input polarisation is linear, aligned to bisect the principal axes of this fibre, then the ellipticity of the polarisation at the top of the fibre depends on the voltage. The light then passes into a section of normal 'spun' fibre in which the current passing along the busbar produces Faraday rotation. Finally, a second section of electro-optic fibre — with its axes appropriately aligned — will effectively produce a second measurement of the voltage. By separating out the ellipticity and the orientation of the polarisation vector at the receiver, voltage and current measurements may be performed simultaneously, using only polarisation as the information carrier for both parameters. The polarisation history along each section may be traced using the Poincaré sphere. There are a few circumstances where ambiguities may arise; for instance, if the polarisation at B (Fig. 13.2) is circular, then Faraday rotation will produce no change in the output state, at C. However, this will, by simple symmetry arguments, appear at D as linear polarisation orthogonal to that input at A, and so this condition may be readily detected. Again, a choice of the material for the fibre with due attention to the anticipated total potential differences will ensure that this condition is avoided. For similar reasons, a pure circular polarisation at D is also difficult to analyse. There is, at present, a snag. A suitable electro-optic material from which to fabricate the fibre has yet to be found. The constraints are difficult. It should be single crystal, yet compatible with being drawn into a fibre form. It should also splice readily with conventional silica fibres which may be used at the current measuring section. But there are very many possibilities, including the use of electrically active dopants in otherwise conventional fibres. Other possibilities include the integrated optic sections in the fibre length to perform the modulation by sampling the field at specific points.

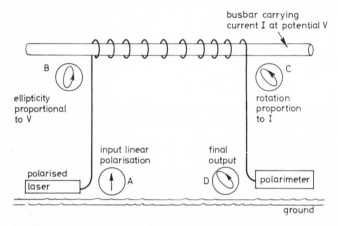

Fig. 13.2 *An all-fibre system for the simultaneous measurement of current and voltage in power lines. Two different types of fibre are involved – an electro-optic section to measure voltage, and one optimised for Faraday rotation to measure the current*

The potential gains of such a system are significant since the optical system is superior to the currently used transformer-based system in ease of installation, cost, accuracy and stability. It is, of course, realistic to consider an alternative system based on a fibre coated with a polymer film of piezoelectric material to phase modulate light in proportion to the total voltage, whilst rotating the plane of polarisation via several turns around the conductor to monitor the current. These two measurements are readily separated, and are accessible to present technology.

Many other multimeasurand configurations are feasible, remembering the intensity, phase, frequency, colour and polarisation of light may all be modulated by the measurand. Clearly, the two or more properties to be modulated must be chosen such that they may be transmitted to and from the measuring region without significant crosstalk, and the modulation processes themselves should be effected with minimum crosstalk. These conditions can certainly be met by appropriate care in conceptual design, optical and mechanical engineering and materials science. This is, however, an area in which little work has been reported.

Multiplexing systems have received much theoretical attention, though again little has been reported experimentally. Numerous sensors can be time or frequency division multiplexed on to a single fibre highway. A typical scheme is shown in Fig. 13.3. The sensors, which in this discussion are intensity modulators – though they could in principle be any of the others with appropriate detection optics – are fed from a pulsed light source. A train of reflected pulses returns to the receiver,

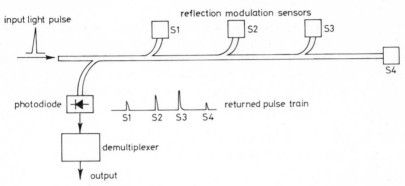

Fig. 13.3   *A sensor multiplexing system for monitoring reflection modulated transducers coupled to a common feed and return fibre path*

each pulse return corresponding to the reflected signal from a particular sensor. If we return to the signal to noise estimates mentioned earlier, we find that a total loss of signal from each sensor of about 50 dB is permissible before the per sensor resolution becomes worse than 0·1%. Effective losses accumulate very rapidly in this system. In many cases, the light source output will be peak power rather than average power limited, so that the peak pulse power will be very close to the actual CW power which would be available for a single-sensor system. If the pulse length is τ, then the pulse separation at the receiver should also be at least the order of τ, and for N sensors the interval between pulses is then of the order of $2N\tau$. The total optical power launched is therefore about $1/2N$ of that available for the single element. Then this power is to be divided between the N sensors, so that the maximum power available per sensor is of the order of $1/2N^2$. To this we must add interconnection losses at each transducer, and allow for the fact that the return power also suffers a loss at each coupler (this will be small provided that the coupling coefficients of each coupler are also small). Occasionally, the actual losses within the fibre transmission path may also be significant. The 50 dB then fairly rapidly disappears. If a connector loss of 1 dB is assumed, then the maximum sensor count is only about ten, and very little of this available bandwidth has been used. At a connector loss of 0·1 dB the total possible count is in the region of 50–100. A similar system may be configured based upon a chirp modulation of the light source intensity waveform. This effectively removes the source modulation loss involved in the pulse system, and replaces the pulse train reciever by a filter bank. Frequencies corresponding to each sensor are obtained by beating the transmitted chirp with the returned signal. A chirp modulation is a linearly advancing frequency modulation, which may be written as:

$$E(t) = e^{j(\omega + \alpha t)t} \qquad (13.1)$$

When this is mixed with a delayed version of itself, the result is:

$$E(t)E^*(t-\tau) = e^{j(\omega + \alpha t)t}e^{-j|\omega + \alpha(t-\tau)|}$$
$$= e^{-j\alpha rt} \qquad (13.2)$$

In practice the frequency ramp cannot continue for ever, and is usually periodic. This complicates the spectrum somewhat, but the general principles remain [13.13]. In the multiplexed sensor case, the amplitude of the output from the filter at frequency $\alpha\tau$ is proportional to the reading from the sensor spaced at τ from the reference signal. For 1 dB connectors, we now lose our 50 dB at about 18–20 sensors in the highway, and well over 100 are feasible on the 0·1 dB per connector highway.

There are at least three other multiplexing systems which have been suggested. Wavelength division multiplexing is a relatively common technique in PTT systems.

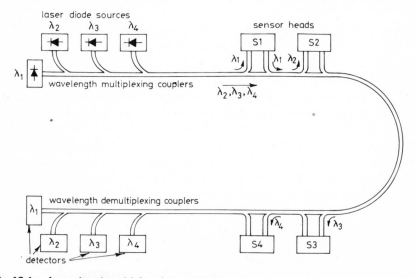

Fig. 13.4   *A wavelength multiplexed sensor highway system*

The basic principles of a sensor system are shown in Fig. 13.4. Many of the necessary components will be developed as a part of the PTT effort, but is should perhaps be mentioned that the emphasis for PTT is quite different. The basic aim in PTT is to optimise the use of the relatively expensive communications medium as a point to point link. Separate optical sources are used because of the difficulties of nonlinearities and crosstalk should one signal source be modulated. The maximum desired channel count is typically about four, representing three outgoing and one return video channel. In the case of the sensor system, more channels would be desirable, though the support technology for more wavelength multiplexed channels would be difficult to achieve. Perhaps one could put the case for combined wavelenghth and time multiplexed channel. A relatively simple system with four sources, mixed in with a wavelength multiplexing loss of 3 dB could then, in principle, support four times as many sensors as a single time multiplexed highway. However, we still have to add the assumed connector losses, and also some multiplexing loss at each sensor input, so the overall picture then looks slightly less optimistic. Clearly, detailed estimates of the performance of such a system depends on the characteristics of the individual components, especially connectors.

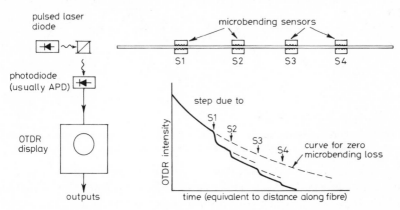

Fig. 13.5   *A microbend sensor highway using optical time delay reflectometry (OTDR) to monitor the sensors*

An interesting variation on this involves the use of backscattered radiation in an optical time domain reflectometer [13.14, 13.15]. This is based upon microbending sensors, and a localised step in the loss may be detected in the returned pulse (Fig. 13.5), whose amplitude is displayed as a function of time. This is attractive in that there are no optical connectors necessary in the sensor highway, but the return losses in OTDR are quite high, and only of the order of 100 resolvable points are available in the entire amplitude spectrum occupied by the total sensor highway. It is therefore a concept which is somewhat limited in overall responses, though these resolution points may be arbitrarily shared between the sensors by appropriate design. There are also complications due to the fact that the modal spectrum is modified at each sensor, so that the output may fluctuate unless some precautions are taken to alleviate this. The concept has also been demonstrated in single-mode fibres using a mode stripper at each sensor to ensure that none of the radiation converted to radiating modes at the sensor is reconverted to guided modes.

Yet another variation, this time using single-mode fibres and integrated optic sensors, is shown in Fig. 13.6 [13.16]. The system is again based on chirp modulation, but this time at an optical frequency. The sensors are interferometric, and so the output from each sensor is a frequency which is uniquely related to the path difference on each interferometer in the chain. By interconnecting both outputs from each interferometer into both inputs the power losses may be minimised, and again the per connector loss is the parameter which limits the maximum total number of sensors in the data highway.

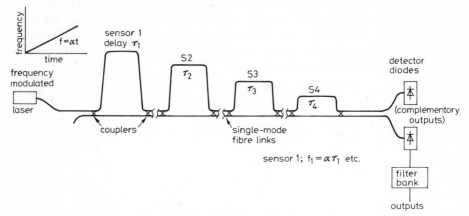

Fig. 13.6 *An interferometric sensor multiplexing highway based on a chirp modulated optical source and differential interferometer delays in each sensor head*

This then has given some indication of some of the proposals for multiplexed optical sensor highways. This is one of many areas which require a full characterisation of the true engineering potential of fibre sensors. There are numerous other potential developments. These include the possibility of passive 'smart' sensors (see Section 13.5) and improvements in the basic fibre transduction processes to permit direct modulation of light within the fibre without going through an intermediate stage. There is also significant potential in the refinement of new processes to optimise the interaction between relatively standard optical measurement techniques and an optical fibre feed and return for the light to energise the measurement process.

Fibre materials can play an important part in future optical transducer systems. There are many areas in which crystalline fibres could be useful. The electro-optic application has already been mentioned. An optical fibre with a defined crystalline structure is clearly a difficult technological proposition, and it may be that defects in the crystal could be responsible for catastrophic mechanical failure, therby turning a difficult problem into a fundamentally impossible one. But the potential of crystalline fibres in sensors, not only electro-optically, but in directionally sensitive strain and pressure sensors and in temperature measurement exploiting polarisation properties, is impressive, and certainly merits a thorough investigation of the required materials technology. The use of doped glasses to render them sensitive to a particular environmental parameter and less sensitive to others is another area in which materials technology can play an important part. A phase modulation magnetic field sensor using a metallic glass which is magnetostrictive has already been demonstrated [13.16]. Similarly, a wide variety of fibre coating techniques have been proposed and tested to enhance the phase sensitivity of a fibre to a particular parameter while reducing sensitivity to others. Coating materials include both metals and plastics. The former result in a more rigid, and the latter in a more compliant, composite structure. The principles of coatings are simple. The coating responds to the measurand field and transmits the response to the fibre via a shear stress (primarily) across the interface between the coating and the secondary coating on the fibre. The mechanical properties of the whole are then determined by the combination of the properties of the fibre and the coatings, and a full mechanical analysis is therefore a little complex [13.17].

This, then, has briefly looked into some of the possible developments in, primarily, low-cost, relatively high-volume industrial sensors and sensor systems. There are a few fibre sensors which are more specialised and warrant individual mention. The major ones here are the fibre gyroscope and the fibre phase modulated hydrophone. The attraction of the hydrophone lies in its geometrical flexibility and in its high sensitivity. The former allows for a readily designed wide variety of polar responses and frequency responses, and the latter, along with suitable correlation signal processing, offers the potential of detecting the presence of coherent signals submerged in sea noise. These are proven possibilities.

Another — as yet unproven — possibility with the fibre optical phase modulated hydrophone is that of a single monomode fibre feed through the hull of a submarine to an array of hydrophones. The hydrophone signals could be processed optically using integrated optic switchable delays, giving both beam steering and combining functions. The array would then be fed with a single coded electrical signal to switch the delay modulators, and via a single optical fibre. The required watertight gland structure through the hull would then be considerably simplified. A schematic of this system is shown in Fig. 13.7. There are no fundamental reasons preventing the fabrication of such a system. The conceptual bases of fibre optic sensors need to develop to attack the lead sensitivity problem, and the interface between integrated optics and single-mode fibres must be made into a routing operation. Numerous approaches are now under evaluation, and a system of this nature will become feasible by the mid 1980s.

The fibre gyroscope is another sensor with specialist applications, and its future role will probably become clear by the mid 1980s. Inertial grade performance has been demonstrated, but its position with respect to competing gyroscope technologies is difficult to asses with certainty. There is a considerable current view that the device may be best suited to rate applications (sensitivities of the order of 1 degree per hour), but here cost is paramount, and numerous novel assembly techniques for mechanical gyroscopes — based on moulded plastics — have succeeded in producing very economic mechanical devices with the required performance. Viewpoints on the required technologies for practical fibre gyroscopes also differ considerably. An all-fibre construction is attractive in that it removes all interfaces and eases the problem of retaining reciprocity. However, integrated optics offers advantages in the availability of simple and flexible modulators and splitter/combiner assemblies. Both approaches require a thorough evaluation, with due attention to the required applications. The remaining unsolved problem with the fibre gyroscope is that of scale factor stability. This will occupy a significant amount of effort – as indeed was also the case with the ring laser gyroscope. A number of approaches are slowly emerging, but it is currently too early to comment on their relative merits.

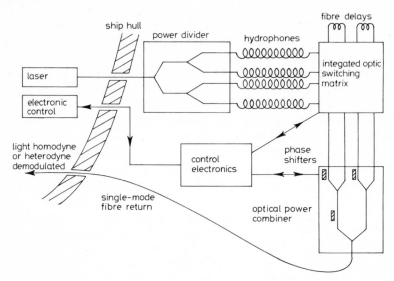

Fig. 13.7   *Schematic diagram of a multiplexed optical fibre hydrophone system incorporating beam forming and combining in the optical domain*

13.4 Optical fibre signal processing systems

Optical fibre signal processing appears to offer a great potential, particularly in delay line applications. The other principal areas – nonlinear devices and spatial processors – seem less promising, at least in the near future. Nonlinear effects are still only observed at very high power densities (GW/cm^2), and even in single-mode optical fibre this typically involves power levels of the order of one watt. This seems to indicate that significant attention has to be paid to the materials problem before the power consumption of nonlinear optical devices becomes compatible with digital processing applications. The required hysteresis effects to produce memory elements switches have, of course, been demonstrated.

The full potential of optical spatial processors remains elusive. Fibre optics could play a part here in the transfer of spatially coded information. This has been done with fibre bundles to good effect, but another possible technique is to use individual modes in a multimode fibre as separate channels. Mode mixing and mode conversion will introduce some crosstalk, but the levels of this, for short transmission distances, could be sufficiently low for successful transmission of digital information. The choice of the modes to be launched is important here, since mode mixing and mode conversion will occur mainly between specific portions of the mode spectrum. The viability of this approach has been demonstrated for short distance transmission (a few metres) over a low-moded fibre.

Optical fibre delay line processors offer an impressive potential, in terms of bandwidth and total available delay. The current devices are limited either by the use of real, positive — and difficult to control — taps, or, in the case of the phase modulated system, by the frequency response of the phase modulators. There are thus two obvious aspects of fibre optic tapped delay lines which could form the basis of future expansion in the area. The first is in the design of high-frequency phase modulators, and the second is in the design of complex taps. This, of course, assumes that the delay line is to operate with a coherent optical carrier. With an incoherent carrier, positive and negative taps would be instrumental in increasing the flexibility of the system considerably, but complex taps — which imply phase information — are excluded unless some form of coherence is introduced into the system. It is feasible to represent complex numbers as incoherent signals, but it is less obvious how the number thus represented may be combined with another complex number.

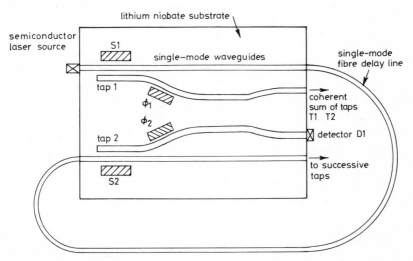

Fig. 13.8   *An integrated optic tapping device for coherent (that is, complex) tapping of an optical fibre delay line*

A high-speed optical phase modulator — certainly with the bandwidths of interest — is readily available in integrated optic form. An electrode on a guide in a lithium niobate substrate immediately fulfils the requirement. Gigahertz bandwidths should be realisable, though there are a number of potential hazards with spurious amplitude modulation caused by the details of the electrode configuration on the guide. Directional coupler switches with comparable frequency responses are also feasible with standard integrated optics technology. Integrated optics technology could also form the basis of a complex optical tap for a fibre optic guide. The general features of such a system are shown in Fig. 13.8. Here a simple two-tap delay line is shown, and the amplitudes and phase of the two taps may be modified and controlled. The taps themselves are two directional coupler switches, and the amplitudes of the taps may be varied by adjusting the bias on S1 and S2 in the diagram. The phase of the taps may be modified by adjusting the voltages on the phase electrodes ϕ_1 and ϕ_2. The output port contains the appropriately phased sum of the two taps. The diode D1 produces a signal which, in the absence of any modulation on the light in the main delay line, may be used to calibrate the relative phases of the taps T1 and T2. Clearly numerous taps may be cascaded to produce a fully coherent optical fibre delay line processor.

It is interesting to compare the potential of this delay line with that of the (now well understood) surface acoustic wave equivalent. The SAW device has the potential of complex taps, and also has the potential of preprocessing an input signal as it is applied to the delay line by varying the electrode finger pattern [13.19]. The SAW device does, however, operate on a compact substrate in which the delays — since they are at acoustic velocities — are relatively long per unit device length. (There is a factor of 10^5 between optical and acoustic velocities.) In the fibre optic case it is certainly possible, by using integrated optic modulators, to apply weighted input data. The delays require the use of fibre rather than integrated sections, unless the anticipated data rates are extremely fast. We then again return to the interface between fibre and integrated guides as an important parameter and, particularly to ensure complex taps, the phase of the delay must be stabilized. Over a delay of the order of centimetres this is difficult, since the temperature coefficient of a fibre guide is of the order of 100 radians per metre per degree. However, a combination of reasonable temperature stabilization and a suitable resetting procedure for the tap phases (perhaps occupying ten microseconds every millisecond, for example) could overcome these difficulties. In common with most of these concepts, a complete evaluation of the potential of fibre optic tapped delay line processors is indicated, and, of the signal processing themes, the fibre delay line appears to be the most promising.

13.5 Hybrid systems

The availability of all the preceding techniques for the manipulation of information in the optical domain leads to the question of what could be done with optical systems involving two or all of the sensing, information processing and data transmission. It is interesting to briefly discuss some of the possibilities.

There is currently a considerable interest in 'smart' sensors in which some electronic signal processing is incorporated into the sensor head. This processing will typically include some provision for self-calibration to compensate for thermal and temporal variations in the sensor characteristics, and often an analogue to digital convertor to facilitate sending the information from the sensor to the control room in a digital form. Could the same concept be implemented optically, but with zero electrical power at the sensor and with long-distance — several kilometres — monitoring of the state of the sensor?

The answer is, of course, in the affirmative, and Fig. 13.9 shows a system which could be manufactured to produce a digital read-out of the position of a moving object — perhaps a diaphragm — relative to some mechanical reference. A typical requirement may be to produce a resolution of ten bits over a total displacement of 25 microns. The reading mechanism is interferometric and uses holograms to transform relative positions into diffraction angles, and the diffraction angles are then coded as bits in a digital number. Clearly, to obtain this resolution, it seems likely that grating manufacture should be implemented to this order of accuracy.

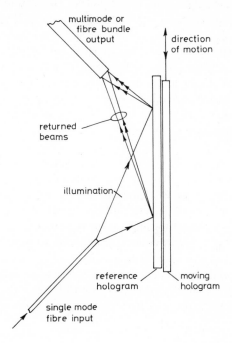

Fig. 13.9   *A conceptual diagram of a holographic system to translate displacement into a digital code*

The holograms would therefore perhaps be in the form of modified Moiré fringe systems. The gratings could then be produced by some form of selective etching technique on crystal planes and the dimensions of the grating defined by the lattice constants, since the required resolution is better than that obtainable using optical photolithography. As a further refinement, the hologram coding could be modified to produce a code which launches each bit of the binary number into a different mode in a multimode fibre. The modes could then be unscrambled at the receiver using an all-optical technique. Perhaps this is a little fanciful. It certainly involves putting laboratory technologies (some yet to be refined) into an industrial environment. But it is no more difficult than putting a computer on to a small chip of silicon. Though, admittedly, the latter has involved millions of man-years of effort. Specialist optical devices cannot — and should not — attract this manpower investment. The tools exist in the laboratory, and they should be evaluated in a realistic framework for these more demanding applications. And simpler devices with coarser resolution (ten microns) have been built [13.20].

This is one example of a family of similar devices which could be proposed to exploit the combination of optical spatial processing and fibre sensing. This particular interface is one which has received little research effort — perhaps more should be known about it. We have already seen some examples of optical fibre delay line processing in sensors — in the serially multiplexed passive sensor highways. As for nonlinear optics, it is currently (1982) premature to comment on the next twenty years of progress, but without a highly sensitive nonlinear material any nonlinear device would immediately counter all hazardous area regulations owing to the optical power densities involved. This somehow defeats the overall objective.

It does, however, appear that there is considerable scope for the passive 'smart' sensor, and the technologies and conceptual framework both exist to make these devices a reality. The self-correcting analogue to digital convertor is particularly attractive, but it will be some time before the practicalities of the technologies involved are characterised with sufficient confidence to give a realistic assessment of their importance.

Sensor networks can readily be used for data transmission as well, and there are countless techniques whereby this may be put into operation. Any system based on coherent light is readily adapted for use with the Fibredyne [13.21] data highway, provided that suitable frequency division demultiplexing is made available at the receiver. Any of the other sensor highways may also be broken into to provide a data channel. Whether a system designer would wish to implement such a combination is unclear, but that he could is interesting.

This list could lengthen infinitely, but it is probably superfluous to do so. The principle purpose of this short section has been to indicate that there is the scope for a blending of technologies, and that the benefits from doing so may be significant.

13.6 Some final comments

This short volume has tried to convey the essential principles and the bases of the fundamental limitations of optical techniques applied to the measurement of physical variables, with particular emphasis on systems involving fibre optics as the transmission medium. The hope is that these fundamental principles will, at least, remain relatively time invarient.

The technology of optical systems is developing rapidly on all fronts — optical sources, accurate modulation and detection techniques, fibre transmission and materials for the fabrication of optical components. All these developments will have some impact on the applications in sensing and signal processing which are the main theme of the book. One of the principal problems will be tracing the appropriate technological advances and filtering out the remainder from the bewildering mass of literature which is appearing on the subject. Maybe this work will have indicated some of the relevant subject matter, but it would be – to say the least – presumptuous to assume that we have discussed all relevant matters in optics in the previous pages.

The final success of optical techniques will depend on a complex mix of opportunities, determination and sheer luck for both the research and development engineer and the eventual manufacturer and user. There is, without doubt, a role for optics in the future for sensing and signal processing systems. Whether it will be relatively limited to the domain of hazardous or remote plant, regions of high electromagnetic interference and similar areas – or to very high rate but simply manipulated data systems – depends on the interaction between a variety of technological and conceptual developments. The answer will be given – in time.

Appendixes

These appendixes have been included to review very briefly most of the basic optical concepts which are relevant in the science of optical fibre sensing and signal processing systems. The descriptions included are intended to provide a sketchy physical interpretation of the formulations mentioned and to give references to source literature, in which more complete descriptions of these basic ideas may be found. To cover all the material with mathematical rigour and providing full physical descriptions would have resulted in an optics text — of which there are several excellent examples already available (see References A.1–A.4, for example).

Appendix 1 Geometrical optics

One can go far in geometrical optics with little more than the thin lens law:

$$\frac{1}{u} + \frac{1}{v} = \frac{1}{f} \tag{A.1}$$

and the observations that wavefronts and rays are normal to each other. Fig. A.1 illustrates the general principles. Application of the principles in Fig. A.1 to the case of the imaging of a source which emits light into a given numerical aperture soon demonstrates that the product of linear dimension and numerical aperture is a constant in an imaging system (for small numerical apertures). It should be noted that the lens law applies to imaging of spatially incoherent sources. Imaging of spatially coherent sources (most commonly laser beams in this context) is dictated by Gaussian beam optics (see Appendix 4) which tends to geometrical optics under certain circumstances, but by no means generally.

The lens formula also applies to imaging with spherical mirrors, again in the paraxial limit — that is, when rays from the object are all within angles with the optic axis defined by the $\sin \theta = \theta$ approximation. In the case of mirrors the focal length is half the radius of curvature, and there are the usual sign conventions for concave and convex surfaces.

The imaging relationships are very useful, and can be used to define the operation of almost all incoherent intensity sensing devices. The sensing mechanism is invariably the defocusing of an image or the interruption of an (often expanded) beam by some form of shutter mechanism. Under most circumstances the roles of both diffraction and abberations are relatively minor, and very simple geometrical ray tracing analyses may be used to good effect.

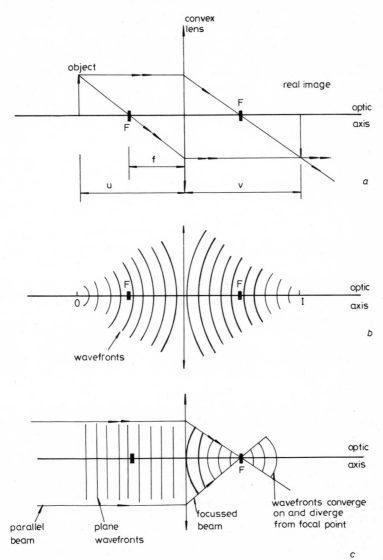

Fig. A.1 *Geometric optics image formation with a convex lens. (a) Ray construction for object and image positions. (b) Wavefront representation of the imaging of an on-axis point of the object in (a). (c) Geometrical and wavefront representations of the focussing action of the lens*

Appendix 2 Diffraction

The simple diffraction grating is well known (Fig. A.2). Constructive interference occurs at angles where:

$$d \sin \theta = \lambda n \tag{A.2}$$

and for small angles, again, $d\theta \sim \lambda n$. The quantity $1/d$ is the fundamental spatial frequency of the grating in lines per unit length and, clearly, n/d represents spatial harmonics of this frequency. If the optical amplitudes in the far field from the diffraction gratings are examined, it can be shown (see for instance Reference A.5) that the amplitudes of the components at the angle $\theta \sim n\lambda/d$ are proportional to the amplitudes of the spatial frequency components at the nth harmonic of $1/d$.

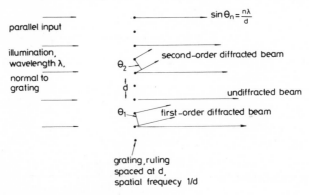

Fig. A.2 *Diffraction from a ruled grating*

This relationship may be generalised to show that the far field pattern is the Fourier transform of the pattern which is illuminated, provided that the illumination is a monochromatic parallel beam along the optic axis of the system.

If the light which is diffracted by the object is collected using a convex lens, then the far field pattern (effectively an object at infinity) is imaged in the back focal plane of the lens. If the original object causing the diffraction is in the front focal plane, then the optical amplitude in the back focal plane is an exact Fourier transform of the object. If the input plane deviates from the front focal plane then there is some wavefront curvature in the back focal plane, resulting in a phase error in the Fourier components. The intensity in the back focal plane is then the power spectrum of the input (see Fig. A.3).

One area in which diffraction plays a central part in optical systems is in defining the minimum spot size which may be produced from a laser beam. Typically, laser beams are Gaussian, with $1/e$ points at a radius w. A Gaussian amplitude distribution will transform to a Gaussian frequency spectrum of width, again, at the $1/e$ point of $2/w$. This then gives for the focused spot size a value of the order of $2 \lambda/w$. Again, though, due caution is required in interpreting this result exactly, since to be strictly correct the Gaussian beam optical analysis should be used. Our estimate is of the correct order if the divergence of the input beam is negligible. This may be the case with gas laser sources, but is not with semiconductor lasers. This diffraction-limited spot may often be achieved in monochromatic optics, but effects of abberations must always be considered even though these may turn out to be small.

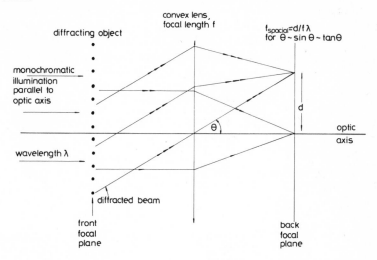

Fig. A.3 *Focusing of diffracted beams by a convex lens to form the spatial Fourier transform in the back focal plane*

Appendix 3 Coherence and interferometry

The topic of the coherence of optical radiation is usually discussed in the context of interferometry – in part to give physical interpretation and means of quantifying the concepts of coherence. Detailed accounts of coherence phenomena may be found in References A.6 and A.7, an introductory account in Reference A.2, and a useful descriptive account in unit 5 of Reference A.3. The Michelson interferometer (see Fig. A.4) is an interferometric configuration which may be used to facilitate the study of temporal coherence effects. Temporal coherence may be defined in terms of the time for which the radiation emitted by a source remains predictable in phase. Assuming that the two interfering beams in the Michelson are of equal amplitudes E, and are caused to interfere spatially perfectly at the detector D (this is an alternative way of say that the two beams overlap exactly, or alternatively that the two beams originate from the same spatial source – that is, are spatially coherent), the total field E_t at D is:

$$E_t = E_1(t) + E_2(t - \Delta t) \qquad (A.3)$$

where $\Delta t = \Delta l/c.$

The intensity at D is therefore:

$$I = E_t E_t^* = E_1 E_1^* + E_2 E_2^* + E_1(t)E_2^*(t - \Delta t) + E_1^* E_2(t - \Delta t) \quad \text{(A.4)}$$

where $*$ denotes taking the complex conjugate. The last two terms in the above expression represent the interference between the two beams. If the optical field is perfectly sinusoidal, then this expression rapidly reduces to the standard interference formulation:

$$I = I_0(1 + \cos \phi) \quad \text{(A.5)}$$

where ϕ is the optical phase difference between the two beams.

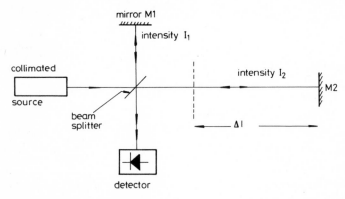

Fig. A.4   *The Michelson interferometer*

However, if the two interfering beams cannot be approximated by a sinusoidal source — that is, if the two beams after time Δt are no longer almost totally temporally coherent — then eqn. A.4 must be evaluated for the temporal properties of the radiation field concerned.

The nature of the observed total result depends on the observation interval τ over which the results of eqn. A.4 are observed. Thus, we may express this as:

$$I(\Delta t)/_\tau = I_1 + I_2 + \langle E_1(t)E_2^*(t - \Delta t) + E_1^*(t)E_2(t - \Delta t) \rangle_\tau \quad \text{(A.6)}$$

where $\langle \rangle$ denotes taking the average over the time τ. If τ is sufficiently long, then the averaging process takes the autocorrelation function of the field with itself delayed by a time Δt. The autocorrelation function is the Fourier transform of the power spectrum of the source (see for instance Reference A.8). We are thus now in a position to relate the power spectrum of a source to the interference produced by that source in a spatially coherent, temporally incoherent interferometer. To illustrate the concepts, the output from such a device for a number of different optical sources is shown in Fig. A.5. Fig. A5*a* shows the normal cosinusoidal fringes observed for a perfectly monochromatic source. Fig. A.5*b* illustrates the effect of a finite linewidth on the fringes observed from a single frequency source. Note that the fringe contrast C, defined by:

$$C = \frac{I_{\max} - I_{\min}}{I_{\max} + I_{\min}} \qquad (A.7)$$

is the autocorrelation function (normalised to the value at zero path difference)

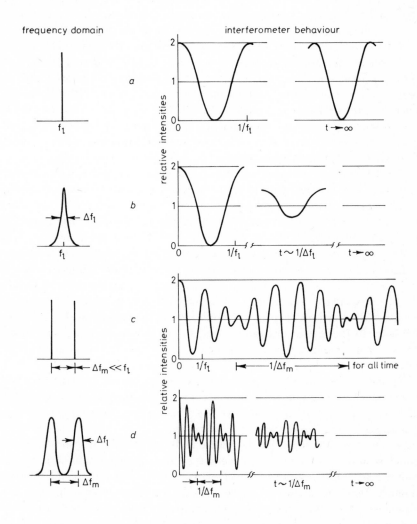

Fig. A.5 *The relationship between frequency domain representation of an optical source and the output from the interferometer as a function of path difference. (a) Single-line source, zero linewidth (b) single-line source, finite linewidth (c) double lines, zero linewidth and (d) double lines, finite linewidth*

of the optical spectrum. This function is also often referred to as the coherence function of the source. The sketches in Figs. A.5c and A.5d generalise the concept for sources emitting at two frequencies (as may be observed in multimode laser, for example). For sources with three or more oscillating frequencies, the situation is somewhat complicated by the effects of dispersion between the modes (that is, slight differences in the frequency difference between adjacent oscillating modes). But using standard Fourier transform rules (see Reference A.8) can always give the autocorrelation function of the spectrum. Obviously, the reverse process may be applied on the contrast function to give the power spectrum.

This introduces some of the formalism of temporal coherence. The references cited will provide a more complete description. Finally, it should be emphasised that the fringe measurement technique for determining temporal coherence properties of a source is only valid if the two interfering sources (that is, the source splitting and recombining processes in the interferometer) are spatially fully matched (or are spatially totally coherent). In performing experiments on coherence effects, this consideration is the one most frequently forgotten when interpreting the results of fringe contrast measurements.

Much is said about coherence times and coherence lengths (related by the velocity of light in the medium under consideration). There is no standard definition of coherence time, though a convenient number, like the contrast function dropping to one half, or $1/e$, is often used. This discussion will also have demonstrated that significant interference effects may occur well beyond the coherence length, particularly concerning effects on noise floors in interferometer devices, where the optical shot noise level may be 10^{10} or more below the carrier. Interference effects may then be significant when the coherence function is of the order 10^{-10} or less! By the same token, deviations of the coherence function from unity by the same amount will effectively increase the noise floor of a detection process. It is beyond the scope of the present discussion to expand further on this topic, but its importance is well recognised in all interferometric systems, which should endeavour to operate over zero path difference to ensure optimum signal to noise ratios.

Spatial coherence may be though of as bearing the same relationship to the spatial spectrum of a source as temporal coherence bears to the temporal spectrum: each is a measure of how close to single-frequency approximations the source characteristics may be defined. The spatial frequency analysis of a source involves expanding the radiation from the source in terms of a set of plane waves [A.5]. Another interpretation of spatial coherence is as an expression of how close a source is to a perfect point source. These two are equivalent through a convex lens (Fourier) transformation, as shown in Fig. A.6. The classical interferometric experiment to demonstrate spatial coherence is the Young's slit illuminated by an extended quasi-monochromatic source, shown in Fig. A.7. (A quasi-monochromatic source is one for which the bandwidth $\Delta\nu$ is much less than the central frequency ν). Fringes appear at the point P, and the contrast of these fringes depends on the separation of the slits P_1 and P_2. If λ is the wavelength of the source then, for fringes to be observed at P, the path difference between AP_1P and BP_2P must be less than, or approximately, the wavelength λ. This is:

$$\theta \Delta l \lesssim \lambda \tag{A.8}$$

Therefore, if P_1 is taken as an origin, fringes will be observed at P if the second pinhole P_2 is within a circular area known as the coherence area A_c, given by:

$$A_c \simeq (R\theta)^2 \sim R^2 \left(\frac{\lambda}{\Delta L}\right)^2 \sim (\lambda/\alpha)^2 \tag{A.9}$$

where α is the angle subtended by the source at the plane A.

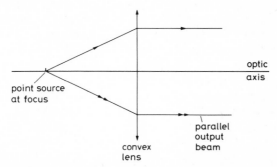

Fig. A.6 *Illustrating the relationship between a point source and a parallel beam in the context of spatial coherence*

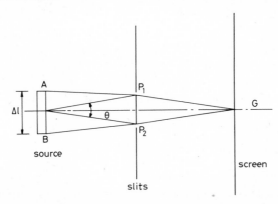

Fig. A.7 *Young's slit experiment with an extended source to illustrate the concepts of spatial coherence*

It is interesting — and highly relevant — to investigate the effects of spatial coherence on the performance of the interferometer in Fig. A.4 using highly coherent sources (both spatially and temporally).

Fig. A.8 shows the situation when one of the mirrors in the interferometer is slightly misaligned, so that the two interfering beams are no longer parallel at the detector. (A similar situation results if the two beams are coaxial but diverge from different effective origins — that is, if the Gaussian beam waists no longer coincide (see Appendix 4). Suppose the phase difference between beams 1 and 2 on the detector is ϕ at point x. The phase difference then clearly varies across the detector face, and may be seen to be:

$$\phi(x) = \frac{x}{w} \left(w \sin \frac{\theta \, 2\pi}{\lambda} \right) + \phi \tag{A.10}$$

Assuming equal beam intensities, the total intensity at a point x on the detector face is:

$$I(x) = (1 + \cos \phi(x)) \tag{A.11}$$

and the total intensity detected is:

$$I_{\text{tot}} = \int_0^w I(x) \propto 1 + \frac{1}{\beta} \{\sin (\beta + \phi) - \sin \beta\} \tag{A.12}$$

where $\beta = \lambda/2\pi w \sin \theta$.

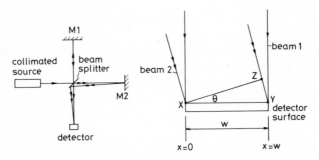

Fig. A.8   *The effects of misalignment of the two beams in an interferometer*

The observed fringe contrast is now given by:

$$C = \frac{I_{\max} - I_{\min}}{I_{\max} + I_{\min}} = |(\sin \beta)/\beta| \tag{A.13}$$

Note that this applies for a source which is otherwise temporally fully coherent, but in which the interfering beams have been spatially misaligned. Similar calculations may be undertaken for different types of spatial beam mismatch. Clearly, severe contrast degradation is inevitable for spatial deviations of greater than one half wavelength — giving required angular alignments of the order $\theta \lesssim \lambda/2w$.

   Thus both spatial and temporal coherence effects play a part in determining the contrast function (and hence the signal amplitude in an interferometric system). As a final note, perhaps this accounts for much of the work on all single-mode fibre interferometers, since here spatial coherence is maintained by the guiding action of the fibre.

Appendix 4 Gaussian beam optics

Gaussian beams are important in two principal contexts — they are the form of the output from most laser devices, and they are very closely related to the spatial modes in parabolic index optical fibres. A full treatment of Gaussian beams may be found in, for instance Reference A.4 or References A.9 and A.10.

A Gaussian beam is usually cylindrically symetrical about the z axis along which it propagates. The electric field corresponding to this beam is given by:

$$E(x,y,z) = E_0 \frac{w_0}{w(z)} \exp\left[-j(kz - \eta(z)) - r^2 \left\{\frac{1}{w^2(z)} + \frac{jk}{2R(z)}\right\}\right]$$

(A.14)

This expression may be interpreted with reference to Fig. A.9. The expression describes a beam of minimum radius w_0 at the origin of the z axis, $z = 0$. The radius w_0, known as the beam waist, is measured to the point at which the field

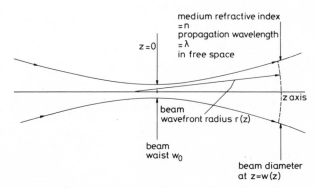

Fig. A.9 *Parameters describing a circular Gaussian beam*

amplitude is $1/e$ of the maximum value. The parameter $w(z)$ describes the evolution of the beam radius as the beam propagates from the origin to the point z. The value of $w(z)$ is given by:

$$w^2(z) = w_0^2 \left(1 + \frac{z^2}{z_0^2}\right)$$

(A.15)

where the parameter z_0 is given as:

$$z_0 = \pi w_0^2 n/\lambda$$

(A.16)

The factor $R(z)$ repressnts the wavefront radius at z, and this is given by:

$$R(z) = z \left(1 + \frac{z_0^2}{z^2}\right)$$

(A.17)

and the factor $\eta(z)$ is a phase correction given by:

$$\eta(z) = \tan^{-1}(z/z_0) \tag{A.18}$$

The beam is cylindrically symmetrical, so that $r^2 = x^2 + y^2$, and k is the wave number $2\pi n/\lambda$.

The parameter z_0 may be interpreted as an expression of the speed with which the beam diverges, and is the distance in which the beam diameter expands by a factor of $\sqrt{2}$. In very rough physical terms, the value of z_0 may be thought of as an equivalent position of a point from which the beam diverges through the beam waist.

The manipulation of Gaussian beams with lenses is important in fibre optics and integrated optics to ensure the match of a laser beam with a guided mode. There is an equivalent to the lens law to describe the transformation of Gaussian beam parameters. The lens law equivalent is:

$$\frac{1}{q_1} - \frac{1}{q_2} = \frac{1}{f} \tag{A.19}$$

where q is the complex beam parameter defined by:

$$\frac{1}{q(z)} = \frac{1}{R(z)} - \frac{j\lambda}{\pi n w^2(z)} \tag{A.20}$$

and q_1 and q_2 are the beam parameters at the input and output sides of the lens. For a thin lens, w is the same at each side, and so we can simplify this to:

$$\frac{1}{R_1} - \frac{1}{R_2} = \frac{1}{f} \tag{A.21}$$

As an example of the application of this, consider the case of focusing a beam from an HeNe laser, which is typically 1 mm in diameter, into a single-mode optical fibre, which supports an (almost) Gaussian mode of diameter 4 microns.

The beam waist is then 0·5 mm and the wavelength is 633 nm, giving for the divergence parameter z_0 1·25 metres. The problem is to determine the focal length of a lens required to focus the beam down to a waist of 2 microns, and to find the position of this waist. The lens law, eqn. A.21, only takes the beam through the lens from the input to output planes. After traversing the lens, we must then propagate the beam according to eqn. A.14. At the waist of the beam the radius parameter $R(z)$ is, by definition, infinite, so that this gives the key to solving the problem. This point we take as a distance p along the light beam. If q_2 describes the beam at plane 2 in Fig. A.10, then $q_3 = q_2 + p$. Using eqn. A.19 — which is simply the result of a matrix representation of geometrical optics (see Reference A.4) — we obtain:

$$\frac{1}{q_2} = \frac{1}{q_1} - \frac{1}{f} = -\frac{1}{f} - j\frac{\lambda}{\pi w_1^2 n} \tag{A.22}$$

Solving for the beam parameter $q_3 = q_2 + p$ gives, after some manipulation:

$$p = \frac{f}{1 + (f/z_{01})^2} \tag{A.23}$$

$$w_z = w_1 \frac{f/z_{01}}{(1 + (f/z_{01})^2)^{1/2}} \tag{A.24}$$

In the example chosen, $w_3/w_1 = 2/500$, giving a focal length requirement of 5 mm and a value of p also of 5 mm. But note carefully that even though this gives values closely related to those which may be expected on an intuitve basis, this is because the input beam chosen has a slow divergence. In the case of a semiconductor laser beam, for example, where the beam divergence is pronounced, a Gaussian beam calculation is always required to ensure that the correct types of lens is specified to perform the transformation. Of course there are practical factors, notably the availability (or otherwise) of the correct lenses, which limit the extent to which the mode matching alluded to here can be achieved in practice.

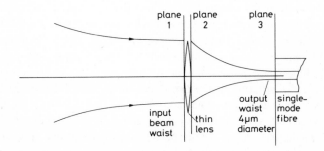

Fig. A.10 *Gaussian optics for lauching a beam into a single-mode optical fibre*

In passing it should be mentioned that the Gaussian beam is in fact the lowest-order mode of a set with similar properties. The higher-order modes have an azimuthal variation, and are most commonly encountered as high spatial output modes (transverse modes) from a laser and as the propagating modes in a parabolic index guide (for instance a lens guide or an optical fibre). It is also possible to manipulate in a similar way with Gaussian beams which are elliptical, the most common being the output spatial modes from a semiconductor laser diode. These beams have different beam parameters in the x and y directions, and require astigmatic optics to perform the transformation to a circular optical fibre guided mode. However, the same procedure will be invoked to design the optical system. Gaussian beams are central to laser systems, and an understanding of their properties is important in the use of the laser as an optical source. More complete treatments of the analysis may be found in References A.4, A.9, A.10 and A.11.

Appendix 5 Polarisation phenomena

Polarisation phenomena are many and varied. Here we shall briefly define various states. We then look at the effects, on a given state, of propagating that state through a specified birefringent path (a birefringent medium is one in which the phase delay through the medium is a function of the polarisation state of the incident light).

Fig. A.11 is a pictorial representation of a number of states of polarisation (SOPs). Linear polarisation (Fig. A.11a) is defined by the direction of the electric vector of the electromagnetic field representing the light propagating along the z axis. Any linear polarisation may be resolved along two perpendicular axes into components travelling in phase along the two axes. If two components on the x and y axes that are equal, linearly polarised but optically 90° out of phase are added together, the result is circular polarised light (Fig. A.11b). The example in the diagram is right circular polarised light — the resultant total polarisation vector rotates clockwise when viewed looking towards the source. Left circular polarised light is generated when the relative phases of E_x and E_y are shifted by 180°. Note that, for circularly polarised light, the orientation of the x and y axes is immaterial; the parameters that specify the light are the fact that it is right or left polarised and its amplitude. Linear polarised light is specified by its orientation and its amplitude.

If the amplitudes of the x and y components which add to form the circular polarised light are no longer equal, then elliptically polarised light results (Fig. A.11c). Elliptically polarised light is fully specified by its amplitude, its ellipticity and the orientation of the major axis of the ellipse with respect to some convenient axes. It is also sometimes convenient to consider that elliptical polarisation is the sum of a circular component and a linear component. The most general SOP is elliptical and off the principal axes of the system (Fig. A.11d). In this case the components on the x and y axes are no longer in quadrature. The total output polarisation may be obtained by resolving one of the components — say that on the y axis — into components which are in phase and in quadrature with the radiation along the x axis. An equal amplitude quadrature component can then be extracted to form circular polarisation, and the remaining component on the y axis added to the in-phase x axis component to form linear polarisation. Finally, these circular and linear components may be added to form off-axis elliptical polarisation. The algebra may be readily inserted into this physical model to give the ellipticity and orientation of the ellipse, and the direction of polarisation rotation.

Birefringence occurs when the refractive index of a material (which must clearly be crystalline, or have molecular handedness and not possess spherical symmetry) is a function of the direction of the electric vector in the lightwave passing through the material. A full calculation of the passage of light through such a material requires the use of the index ellipsoid to find the directions of the so-called ordinary and extraordinary rays. However, in most cases of interest in optical fibres the incident light will propagate along one of the principal axes of the material. The principal axes are those for which the direction of the emerging light ray is independent of the input state of polarisation. If, for instance, the input is directed along the z axis (see Fig. A.12) then there will be two other principal axes (usually orthogonal to each other and the z axis) for which linearly polarised light incident

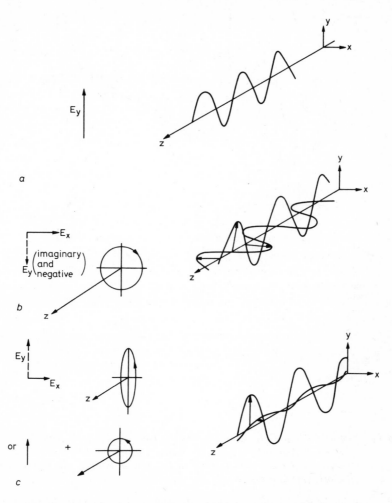

on that axis will emerge linearly polarised on the same axis. These are known as the principal axes of the material, for a linearly birefringent material (that is, a material for which the refractive index is a function of the linear input state of polarisation). The effect of a linearly birefringent material may be calculated by noting the difference in phase for light propagating on the x and y axes, resolving the incident light into components on x and y, then reassembling the components at the output into a defined SOP. The example shown in the diagram is for a quarter-wave plate — in which the phase difference between x and y is 90° —

where linear input polarisation incident at an angle bisecting the principal axes is converted to left circular output. Clearly, adjusting the input orientation through $90°$ gives right circular output polarisation.

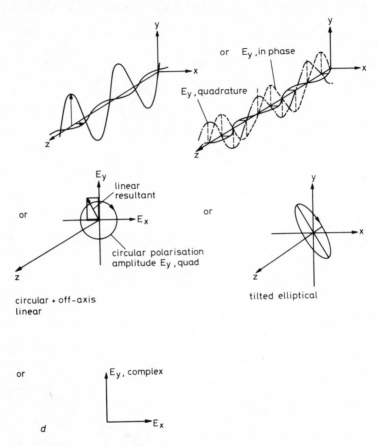

Fig. A.11   *Polarisation states represented as vectors and equivalent travelling waves. (a) linear polarisation on the y axis (b) circular polarisation, here clockwise (or right circular) (c) left elliptical polarisation (d) general elliptical polarisation off the principal axes of the system*

An optically active material is circularly birefringent, so that the input polarisation must be resolved into right and left circularly polarised components, then propagated through the crystal with different optical phases before reassembling at the output. In the former case — linear birefringence — the linear SOP on x and y axes are referred to as the eigenmodes of the polariser, that is, they are the SOPs which will emerge from the crystal unperturbed. In the case of circular polarisation, the eigenmodes are left and right circular polarisation. In both cases the eigenmodes are the ones into which the input radiation must be resolved for passage through the material in order to analyse the total effect.

These modes are orthogonal to each other; that is, if two sources, one in each mode, are brought together and optically interfered, even if they are temporally and spatially coherent, there will be no interference. There is an infinite set of such mode pairs, of which right and left circular, and horizontal and vertical linear SOPs are but two.

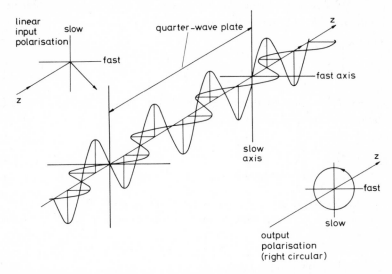

Fig. A.12   *A waveform representation of the action of a quarter-wave plate*

In general, an optical material will be both linearly and circularly birefringent, and one valid way of approaching the analysis is to split the birefringence into two components, and pass the radiation from the linear section into the circular section. The two add linearly, even though in practice the sources of the birefringence may intermingle along the optical path.

There are numerous more formal ways of handling the problem of the trans-mission of polarisation through a birefringent medium. All are in effect a shorthand notation for the general model which we have described — namely, the resolution of the SOP at the input into the appropriate components, the passage of these through a linearly birefringent section, conversion of these components into circular equivalents, then passage through the circular section, followed by addition of these components into an equivalent output SOP. Of these, the most important are probably the Jones vector notation [A.13] and the Poincaré sphere construction [A.14].

The Jones vector notation takes the x and y components and writes them as a column vector

$$E = \begin{pmatrix} E_x \\ E_y \end{pmatrix} \tag{A.25}$$

In this notation, horizontally polarised light of unit amplitude is written as:

$$\begin{bmatrix} 1 \\ 0 \end{bmatrix} \qquad \text{(A.26)}$$

and unit amplitude right circular polarised light, in which the y component lags the x component by $90°$, may be written as:

$$\frac{1}{\sqrt 2}\begin{bmatrix} 1 \\ -j \end{bmatrix} \qquad \text{(A.27)}$$

The extension to other SOPs is straightforward. Passage through a polarising medium may be expressed as a matrix product:

$$\begin{bmatrix} E_{xo} \\ E_{yo} \end{bmatrix} = \begin{bmatrix} a & b \\ c & d \end{bmatrix} \cdot \begin{bmatrix} E_{xin} \\ E_{yin} \end{bmatrix} \qquad \text{(A.28)}$$

where the square matrix represents the effect of the polariser. As an example, if the polariser is a linear polariser with the fast axis horizontal, then the Jones matrix representation is:

$$\begin{bmatrix} E_{xo} \\ E_{yo} \end{bmatrix} = \begin{bmatrix} 1 & 0 \\ 0 & 0 \end{bmatrix} \begin{bmatrix} E_{xin} \\ E_{yin} \end{bmatrix} \qquad \text{(A.29)}$$

A nonpolarising path has unit diagonal elements at a and d, and a quarter-wave plate, with the fast (shorter) axis horizontal, may be expressed by:

$$\begin{bmatrix} E_{xo} \\ E_{xin} \end{bmatrix} = e^{j\pi/4}\begin{bmatrix} 1 & 0 \\ 0 & j \end{bmatrix} \begin{bmatrix} E_{xin} \\ E_{yin} \end{bmatrix} \qquad \text{(A.30)}$$

In a similar fashion, any circular and/or linearly birefringent path can be represented. Successive polarising sections are represented by cascading the matrices – in the correct order – to give a total matrix which operates on the optical incident field.

The Jones matrix representation involves phase information, and is therefore only really suited for dealing with coherent or quasi-coherent light. Totally incoherent light (be it spatially or temporally) should normally be split into quasi-coherent components, since most polarising sections are wavelength selective and directional.

There are other matrix representations of polarisation, including Mueller vectors and Stokes parameters. These are a little more obscure in origins when compared with the Jones vector. The derivation of these, and more details on the Jones notation, are given in References A.2 and A.1. One particularly useful representation of the Stokes parameters is the Poincaré sphere, which is a geometrical construction to permit a simple calculation of the evolution of the SOP of an arbitrary input polarisation through an arbitrarily birefringent medium. Fig. A.13 shows a diagram of the Poincaré sphere and indicates the means by which the sphere may be operated. It is derived from the Stokes parameters, and the equator on the sphere represents states of linear polarisation. An arbitrary origin may be chosen as the x axis, and linear y polarisation is represented by the point y on the sphere diametrically opposite. The north and south poles on the sphere represent circular right and left polarised light respectively. Lines of constant latitude on the sphere describe SOPs of constant ellipticity ϵ given by $\epsilon = \tan \beta$ and the orientation of the ellipse is given by $\beta = \alpha$ with respect to the x axis origin.

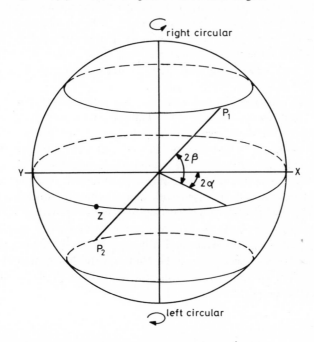

Fig. A.13 *Representation of polarisation modes on the Poincaré sphere*

Any arbitrary birefringent section may be characterised in terms of two quantities, the eigenmodes through that section and the phase difference for propagation of these eigenmodes. Thus, for instance, a quarter-wave plate would be specified by its eigenmodes along x and y axes, and the 90° phase difference between the two eigenmodes. Similarly, an optically active region may be represented in terms of the right and left circular polarised eigenmodes and the phase difference propagation between these two eigenmodes. In general, the eigenmodes (that is, the SOPs which pass through the birefringent section unaltered) are elliptical, but it always transpires that the two eigenmodes may be represented by diametrically opposite points on the Poincaré sphere. Thus, points P_1 and P_2 could represent typical eigenmodes for the birefringent path.

The effect of the passage of a particular SOP through this path is calculated using the sphere in the following manner. The eigenmodes are identified, and then the position of the input state on the sphere is identified. If the phase difference between the two eigenmodes is θ, then the position of the input polarisation on the sphere after rotation of the sphere through the angle θ about the diameter joining the eigenmodes represents the output polarisation. Since any rotation about an arbitrary axis on the sphere may be resolved into polar and equatorial rotations, then any path may be resolved into circularly birefringent components (with eigenmodes on the poles) and linearly birefringent component (with eigenmodes on the equator). As an example, a quarter-wave plate would have the points X and Y as eigenmodes. If the input polarisation to a quarter-wave plate is linear and at 45° to the principal axes — point Z on the sphere — then rotating the sphere by 90° puts point Z on to one or other of the poles, representing circular polarisation. Putting circular polarisation into a quarter-wave plate rotates the pole on to the equator at an angle exactly bisecting the principal axes, so that linear polarisation emerges.

Similarly, a half-wave plate — with the same eigenmodes, but a 180° phase difference — will always produce linear output polarisation from linear input, taking the input to an output which is the reflection of the input in a plane defined by the poles and the X—Y diameter. A half-wave plate always reverses the direction of circularly polarised light, and this is immediately clear from the geometry of the sphere. The general problem of passing monochromatic polarised light through a birefringent lossless path is solved with similar simplicity, once the eigenmodes and phase difference have been defined, and the input polarisation located on the surface of the sphere. There is, however, some contrast with the Jones vector approach, in that the sphere deals implicitly with a lossless path whereas losses — which are polarisation selective — can be built into the Jones matrix formulation. However, the Poincaré sphere is probably the most useful of all aids to the calculation of polarisation phenomena.

References

Chapter 1

1.1 KAO, C. K., and HOCKHAM, G. A.: 'Dielectric fibre surface waveguides for optical frequencies', *Proc. IEE,* 1966, **113**, p. 1151

1.2 DAVIES, D. E. N., and KINGSLEY, S. A.: 'The use of optical fibres as instrumentation transducers', CLEO, San Diego, 26–27 May 1976' 'digest of technical papers' pp. 24–25

1.3 VALI, V., and SHORTHILL, R. W.: 'Fibre ring interferometer', *Appl. Opt.,* May 1976, **15**, pp. 1099–1100

1.4 GOODMAN, J. W.: 'Introduction to Fourier optics' (McGraw-Hill, 1968)

1.5 Information on optical signal processing is covered largely in the appropriate 'SPIE proceedings' (Society of Photo-Optical Instrumentation Engineers, 405 Fieldston Road, Bellingham, Wash. 98227, USA)

1.6 Optical communications is covered by, for instance, the annual 'European conference on optical communications proceedings' (published by the IEE) and the 'International conferences on optics and optical fibre communications (IOOC) proceedings' available through the IEEE and the IEE

1.7 Integrated optics is covered largely through conference proceedings (see [1.6]), the Annual Topical Meeting on Integrated and Guided Wave Optics (arranged through IEEE and the Optical Society of America) and the biannual European Conferences on Integrated Optics (IEE). Also see:
TAYLOR, H. F., and YARIV, A.: 'Guided wave optics', *Proc. IEEE,* 1974, **62**, p. 1044
MARCUSE, D. (Ed.): 'Integrated optics' (IEEE Press, 1972)
TAMIR, T.: 'Integrated optics' (Topics in Applied Physics, Vol. 7, Springer Verlag, 1978)

1.8 Special issue of *IBM Journal of Research and Development*, March 1980, is devoted to Josephson junction devices.

Chapter 2

2.1 MIDWINTER, J. E.: 'Optical fibres for transmission' (Wiley, 1979)

2.2 KAO, C. K. (Ed.): 'Optical fibre technology II' (IEEE Press, 1981)

2.3 See, for instance:
HECHT, E., and ZAJAC, A.: 'Optics' (Addison Wesley, 1973); or
BORN, M., and WOLF, E.: 'Principles of optics' (Pergamon Press, 1974, 5th edn.)

2.4 OKAMOTO, K., HOSAKA, T., and SASAKI, Y.: 'Linearly single polarisation fibres with zero polarisation mode dispersion', *IEEE J. Quantum Electron,* April 1982, **18** (4), p. 496

2.5   KAMINOW, I. P.: 'Polarisation in optical fibres', *ibid.,* January 1981, **17** (1), p. 15

2.6   PERSONIK, S. D.: 'The photon probe: an optical fibre time domain reflectometer', *Bell Syst. Tech. J.,* 1977, **56**, pp. 355–366

2.7   NAGEL, S. R., MacCHESNEY, J. B., and WALKER, K. L.: 'An overview of the modified chemical vapor deposition (MCVD) process and performance', *IEEE QU,* **18**, p. 459

2.8   See, for instance, 'Mid infra-red materials' in Physics for Fibre Optics II (B. Bendow and S. S. Mitra eds) American Ceramic Society 1981

2.9   MALYON, D. J., and SMITH, D. W.: 'The application of injection locked semiconductor lasers for optical communications in the wavelength range 0.8 to 1.6 microns' *in* 'Proceedings comm. '82', *IEE Conf. Publ. 209,* 1983, p. 285

2.10 Component information available in, for instance, the 'IFOC handbook and buyers guide' (Information Gatekeepers)

2.11 BARLOW, A. J., RAMSKOV-HANSEN, J. J., and PAYNE, D. N.: 'Birefringence and polarisation mode dispersion in spun single mode fibres', *Appl. Opt.,* 1981, **20**, p. 2962

2.12 See, for instance:
'Proceedings OFC/CLEO, 1982' (published by IEE and Optical Society of America), 'Proceedings FOC '82' (Information Gatekeepers), 'European optical communications conference proceedings' (IEE) and 'First international conference on optical sensors proceedings' (IEE, 1983).

Chapter 3

3.1   KRESSEL, H. (Ed.): 'Semiconductor devices for optical communications' (Topics in Applied Physics, Vol. 39, Springer Verlag, 1980)

3.2   YARIV, A.: 'Quantum electronics' (Wiley, 1975)

3.3   YARIV, A.: 'Introduction to optical electronics' (Holt, Rinehart and Winston, 1976)

3.4   THOMPSON, G. H. E.: 'Physics of semiconductor laser devices' (Wiley, 1980)

3.5   HOWES, M. J., and MORGAN, D. V. (Eds): 'Optical fibre communications: devices, circuits and systems' (Wiley, 1980)

3.6   MILLER, S. E., and CHYNOWETH, A. G. (Eds): 'Optical fibre telecommunications' (Academic Press, 1979)

3.7   SANDBANK, C. P. (Ed.): 'Optical fibre communication systems' (Wiley, 1980)

3.8   BERGH, A. A., and DEAN, D. J.: 'Light emitting diodes' (Clarendon Press, 1976)

3.9   See, for instance, Unit 5 of Open University Course on Images and Information, course code ST291, (Open University)

3.10 HECHT, E., and ZAJEC, A.: 'Optics (Addison Wesley, 1973)

3.11 FRANCON, M.: 'Optical interferometry' (Academic Press, 1966)

3.12 BORN, M., and WOLF, E.: 'Principles of optics' (Pergamon Press, 1974)

3.13 EPWORTH, R. E.: 'The phenomenon of modal noise in analogue and digital optical fibre systems', 'Technical digest 4th European conference on optical communications', Genoa, Italy, September 1978, p. 492 (IEE)

3.14 GOODMAN, J. W., and RAWSON, E. G.: 'Statistics of modal noise in fibres, a case of constrained speckle', *Opt. Lett.,* July 1981, **6** (7), p. 324

3.15 HARDER, C., KATZ, J., MARGALIT, S., SHACKAM, J., and YARIV, A.: 'Noise equivalent circuit of a semiconductor laser diode', *IEEE J. Quantum Electron.* March 1982, **18** (3), p. 333

3.16 PETERMANN, K., and Weidel, E.: 'Semiconductor laser noise in an interferometer system', *ibid.,* July 1981, **17**, pp. 1251–1256

Chapter 4

4.1 YARIV, A.: 'Introduction to optical electronics' (Holt, Rinehart and Winston, 1976) Chaps 10 and 11

4.2 SZE, S. M.: 'Physics of semiconductor devices' (Wiley, 1981, 2nd edn.)

4.3 KRESSEL, H., and BUTLER, J. K.: 'Semiconductor lasers and heterojunction LEDs' (Academic Press, 1977)

4.4 Technical Staff of CSELT: 'Optical fibre communications' (McGraw-Hill, 1980)

4.5 OSTROWSKY, D. B. (Ed.): 'Fibre and integrated optics' (Plenum Press, 1978) (NATO Advanced Study Institute Series B Physics, Vol. 41)

4.6 Silicon APD data available from many major semiconductor manufacturers

4.7 III-V photodetector diodes available through, for instance, Plessey Microwave and Optoelectronics, Towcester, England

4.8 ANDERSON, L. K., and McMURTY, B. J.: 'High speed photodetectors', *Proc. IEEE,* 1966, 54, p. 1335

4.9 HUBBARD, W. M.: 'Utilisation of optical frequency carriers for low and moderate bandwidth channels', *Bell Syst. Tech. J.,* 1973, 52, pp. 731–765

4.10 WEBB, P. P., McINTYRE, R. J., CONRADI, J.: 'Properties of avalanche photodiodes', *RCA Rev.,* 1974, 35, p. 235

4.11 SMITH, D. R., HOOPER, R. C., and GARRETT, I.: 'Receivers for optical communications, a comparison of avalanche photodiodes with PIN FET hybrids', *Opt. & Quantum Electron.,* 1978, 10, pp. 293–300

4.12 MONHAM, K. L., BURGESS, J. W., and MABBITT, A. W.: 'PIN FET receivers for long wavelength optical fibre systems' *in* 'Proceedings communications '82', *IEE Conf. Publ. 209,* 1983, pp. 280–284

4.13 See, for instance:
YANG, E. S.: 'Fundamentals of semiconductor devices' (McGraw-Hill, 1978; and
SZE, S. M.: 'Physics of semiconductor devices' (Wiley, 1981, 2nd edn.)

4.14 See product data from photomultiplier tube manufacturers, for instance, EMI in the UK

4.15 See, for instance:
SCHWARTZ, M.: 'Information transmission, modulation and noise' (McGraw-Hill, 1980); or
SHANNON, C. E.: 'A mathematical theory of communication', *Bell Syst. Tech. J.,* 1948, 27, p. 379

4.16 STILLMAN, E., and WOLFE, C. M.: 'Avalanche photodiodes' (Semiconductors and semimetals, Vol. 12, Academic Press, 1977)

4.17 TERCH, M. C.: 'Infrared heterodyne detection', *Proc. IEEE*, 1968, 56, p. 37

Chapter 5

5.1 GOODMAN, J. W., and RAWSON, E. G.: 'Statistics of modal noise in optical fibres – a case of constrained speckle', *Opt. Lett.,* July 1981, 6 (7), p. 324

5.2 EPWORTH, R. E.: 'The phenomenon of modal noise in analogue and digital optical fibre systems' *in* 'Proceedings 4th ECOC', Geneva, 1978, p. 492, (IEE)

5.3 CULSHAW, B., BALL, P. R., POND, J. C., and SADLER, A. A.: 'Data collection using optical fibres', *Electron. and Power,* February 1981, 27 (2), p. 148

5.4 BORN, M., and WOLF, E.: 'Principles of optics' (Pergamon Press, 1975) Chap. 7

5.5 YAMAMOTO, Y., and KIMURA, T.: 'Coherent optical fibre transmission systems', *IEEE J. Quantum Electron.,* June 1981, 17 (6), p. 919

5.6 BUCZEC, C. J., and FREIBERG, R. J.: 'Hybrid injection locking of higher power CO_2 lasers', *ibid.,* July 1972, 8 (7), p. 641

5.7   DANBRIDGE, A., TVETEN, A. B., MILES, R. O., and GIALLORENZI, T. G.: 'Laser noise in fibre optic interferometer systems', *Appl. Phys. Lett.*, September 1980, 37, pp. 526–528

5.8   GAILLORENZI, T. G., BUCARO, J. A., DANDRIDGE, A., SIEGEL, G. H., COLE, J. H., RASHLEIGH, S. C., and PRIEST, R. G.: 'Optical fibre sensor technology', *IEEE J. Quantum Electron.*, April 1982, 18 (4), pp. 626–665

5.9   DANDRIDGE, A., TVETEN, A. B., and GIALLORENZI, T. G.: 'Phase noise measurements on six single mode laser diodes' *in* 'Proceedings Conference on Integrated Optics and Optical Fibre Communications', San Francisco, April 1981 (Optical Society of America)
PETERMANN, K., and ARNOLD, G.: 'Noise and distortion characteristics of semiconductor lasers in optical fibre communications systems', *IEEE J. Quantum Electron.*, April 1982, 18, pp. 543–554

5.10 HECHT, A. E., and ZAJAC, A.: 'Optics' (Addison Wesley, 1973) p. 266

5.11 RAMACHANDRAN, G. N., and RAMASESHAM, S., *in* FLÜGGE, S. (Ed.); Handbook of physics', (Springer, 1961) Vol. 25, p. 1; or
ULRICH, R.: 'representation of co-directional coupled waves', *Opt. Lett.*, 1977, 1, p. 109

5.12 BORN, M., and WOLF, E.: 'Principles of optics' (Pergamon Press, 1975) p. 31

5.13 DURST, F., MELLING, A., and WHITELAW, J. H.: 'Principles and practice of laser Doppler anemometry' (Academic Press, 1976)

5.14 FRANCIS, J. J.: 'The design of optical spectrometers' (Chapman and Hall, 1969)

5.15 H. D. Polster, 'Multiple beam interferometry,' *Applied Optics* 8, 552 (1969)

5.16 TOMLINSON, W. J.: 'Wavelength multiplexing in multimode optical fibres', *Appl. Opt.*, 1977, 16, p. 2180

5.17 KOGELNIK, H.: 'Coupled wave theory for thick hologram grating', *Bell Syst. Tech. J.*, 1969, 48, p. 2909, or

Chapter 6

6.1   Gaskill, J. D.: 'Fourier transforms, linear systems and optics' (Wiley, 1978)

6.2   MIDWINTER, J. E.: 'Optical fibres for transmission' (Wiley, 1979) p. 248

6.3   MIDWINTER, J. E.: *ibid.* p. 72 ff and p. 82 ff.

6.4   HARMER, A. L.: 'Displacement, strain and pressure transducers using microbending effects in optical fibres' *in* 'XX general assembly of URSI, Book of Abstracts', p. 383

6.5   LAGALOS, N., MACEDO, P., LITOWITZ, T., MOHR, R., and MEISTER, R.: 'Fibre optic displacement sensor', *in* 'Physics of fibre optics II' (American Ceramic Society, 1981) pp. 539–544

6.6   BLACK, P. W.: British patent

6.7   HARMER, A. L.: 'Principles of optical fibre sensors and instrumentation' *in* 'Proceeding of optical sensors and optical techniques in instrumentation', November 1981 (Institute of Measurement and Control, London, England)

6.8   HECHT, E., and ZAJEC, A.: 'Optics' (Addison Wesley, 1973) p. 79

6.9   PHILLIPS, R. L.: 'Proposed fibre optic acoustic probe', *Opt. Lett.*, July 1980, 5 (7), p. 318

6.10 See for instance: YARIV, A.: 'Introduction to optical electronics' (Holt, Rinehart and Winston, 1976) p. 371

6.11 SHEEM, S. K., and COLE, J. H.: 'Acoustic sensitivity of single mode optical power dividers', *Opt. Lett.*, 1979, 4, pp. 322–324

6.12 LAGAKOS, N., LITOWITZ, T., MACEDO, P., MOHR, R., and MEISTER, R.: 'Multimode optical fibre displacement sensor', *Appl. Opt.*, 1981, **20**, p. 167

6.13 US Patent 4158310

6.14 FROMM, I., and UNTERBERGER, H.: 'Direct modulation of light by sound' *in* 'Proceedings 1978 International Optical Computing Conference', pp. 40–42, London, September 1978 (IEEE, New York, 1978)

6.15 ROBERTSON, C.: Proceedings Electro-Optic Systems Design Conference New York 1969 p. 178 (IEEE)

6.16 See 'Proceedings of optical sensors and optical techniques in instrumentation' (Institute of Measurement and Control, London, November 1978)

6.17 GIALLORENZI, T. G., BUCARO, J. A., DANDRIDGE, A., SIGEL, G. H., COLE, J. H., RASHLEIGH, S. C., and PRIEST, R. C.: 'Optical fibre sensor technology', *IEEE J. Quantum Electron.*, April 1982, **18** (4), pp. 626–666

6.18 'Fibre optics for process control and business communications – transducers and sensors', report no. 82-7, ERA Technology Limited, Leatherhead, Surrey, UK

6.19 MENADIER, C., KISSENGER, C., and ADKINS, H.: 'The Fotonic sensor', *Instrum. & Control Syst.*, 1967, **40**, p. 114

6.20 UK Patent application 2025608

6.21 KYUMA, K., TAI, S., SAWADA, T., and NUNOSHITA, N.: 'Fibre optic instrument for temperature measurement', *IEEE J. Quantum Electron.*, April 1982, **18**, p. 676

Chapter 7

7.1 BORN, M., and WOLF, E.: 'Principles of optics' (Pergamon Press, 1975) Chap. 7

7.2 VALI, V., and SHORTHILL, R. W.: 'Fibre ring interferometer', *Appl. Opt.*, May 1976, **15** (5), pp. 1099–1100

7.3 KINGSLEY, S. A., and DAVIES, D. E. N.: 'The use of optical fibres as instrumentation transducers', CLEO, San Diego, 26–27 May 1976, 'Digest of technical papers', pp. 24–25

7.4 CULSHAW, B., FIDDY, M. A., HALL, T. J., and KINGSLEY, S. A.: 'Fibre optic acoustic sensors', Ultrasonic International, 1979, Graz, May 1979, 'Proceedings' pp. 267–272 (IPC); and
BUCARO, J. A., and COLE, J. H.: 'Acousto-optic sensor development', 'Proceedings EASCON', 1979, IEEE Pub 79CH1476-1, pp. 572–580

7.5 YARIV, A., and WINSOR, H.: 'Proposal for the detection of magnetostrictive perturbation of optical fibres', *Opt. Lett.*, 1980, **5**, pp. 87–89; and
DANDRIDGE, A., TVETEN, A. B., SIGEL, G. H., WEST, E. J., and GIALLORENZI, T. G.: 'Optical fibre magnetic field sensor', *Electron. Lett.*, 1980, **16**, p. 408

7.6 TVETEN, A. B., DANDRIDGE, A., DAVIS, C. M., and GIALLORENZI, T. G.: 'Fibre optic accelerometer', *ibid.*, 1980, **16** (22), p. 854

7.7 BUTTER, C. D., and HOCKER, G. B.: 'Fibre optic strain gauge', *Appl. Opt.* 1978, **17** (18), p. 2867

7.8 CULSHAW, B., DAVIES, D. E. N., and KINGSLEY, S. A.: 'Fibre optic strain pressure and temperature sensors' *in* 'Proceedings 4th ECOC', Genoa, Italy, 1978 (IEI Milan, 1978)

7.9 MIDWINTER, J. E.: 'Optical fibres for transmission' (Wiley, 1979)

7.10 CROSSLEY, S. D.: internal report, University College London, 1982

7.11 KINGSLEY, S. A.: internal report, University College London, 1976

7.12 KINSLER, L. E., and FREY, A. R.: 'Fundamentals of acoustics' (Academic Press, 1962); and HEARN, E. J.: 'Mechanics of materials' (Pergamon, 1977)

7.13 HALL, T. J., and HOWARD, D.: 'Interaction of high frequency sound with fibre-guided coherent light', *Electron. Lett.*, 1978, **14** (19) p. 620

7.14 CULSHAW, B., GILES, I. P., SADLER, A. A., and POND, J. C.: 'The "Fibredyne" data collection system for industrial telemetry applications' *in* 'Proceedings SPIE conference on fibre optics – short and long haul measurements and applications', San Diego, August 1982, 'Proceedings SPIE', Vol. 355, paper 20

7.15 CULSHAW, B., and HUTCHINGS, M. J.: 'An optical fibre flowmeter', *Electron. Lett.*, 1979, 15 (18), p. 569

7.16 CULSHAW, B., DAVIES, D. E. N., and KINGSLEY, S. A.: 'Multimode optical fibre sensors', *in* 'Physics of fibre optics II' (American Ceramic Society, 1981)

7.17 BARRETT, D. J.: M.Sc. thesis, University of London, 1978

7.18 JUNGERMAN, R. L., BOWERS, J. E., GREEN, J. B., and KINO, G. S.: 'Fibre optic laser probe for acoustic wave measurements', *Appl. Phys. Lett.*, February 1982, 40 (4), p. 313

7.19 KOO, K., TRAN, D., and SHEEM, S.: 'Single mode fibre directional couplers fabricated by twist etching techniques' *in* 'Proceedings 3rd international conference on integrated optics and optical fibre communications', San Francisco, April 1981, paper TuJl

7.20 VILLARRUALL, C. A., and MOELLER, R. O.: 'Fused single mode–mode fibre access coupler', *Electron. Lett.*, 1981, 17, p. 243

7.21 BERGH, R. A., KOTLER, G., and SHAW, H. J.: 'Single mode fibre optic directional coupler', *ibid.*, 1980, 16, p. 260

7.22 CULSHAW, B., and WILSON, M. G. F.: 'Integrated optic frequency shifter modulator', *ibid.*, 1981, 17, pp. 135–136

7.23 VOGES, E., OSTWALD, Ọ., SCHIEK, B., and NEYER, A.: 'Optical phase and amplitude measurement by single sideband homodyne detection', *IEEE J. Quantum Electron.*, January 1982, 18 (1), p. 124

7.24 HEISMANN, F., and ULRICH, R.: 'Integrated optical single sideband modulator and phase shifter', *ibid.*, April 1982, 18 (4) p. 767

7.25 COLE, J. H., DANVER, B. A., and BUCARO, J. A.: 'Synthetic heterodyne interferometric demodulation', *ibid.*, April 1982, 18, p. 694

7.26 BALL, P. R., and CULSHAW, B.: 'Digital modulation and phase swept diversity in a coherent multimode fibre system', 'Proceedings 4th ECOC', Genoa, Italy, Sept. 1978, pp. 546–553

7.27 GIALLORENZI, T. G., BUCARO, J. A., DANDRIDGE, A., SIGEL, G. H., RASHLEIGH, S. C., and PRIEST, R. G.: 'Optical fibre sensor technology', *IEEE J. Quantum Electron.* April 1982, 18, p. 626

7.28 EICKOFF, W.: 'Temperature sensing by mode–mode interference in birefringent optical fibres', *Opt. Lett.*, April 1981, 6 (4), p. 204

7.29 LeFEVRE, H. C.: 'Single mode fibre fractional wave devices and polarisation controllers', *Electron. Lett.*, 1980, 16 (17), p. 778

7.30 HALL, T. J.: 'High linearity multimode optical fibre sensor', *ibid.*, 1979, 15, p. 405

7.31 CULSHAW, B., BALL, P. R., POND, J. C., and SADLER, A. A.: 'Optical fibre data collection', *Electron. & Power*, February 1981, 27 (2), p. 148

7.32 MARTINELLI, M.: unpublished work

7.33 KINGSLEY, S. A.: 'Fibre optic gravitational telescope' *in* 'Proceedings first international conference on fibre optic rotation sensing and related technologies', MIT, November 1981 (Springer Verlag, 1982)

7.34 KINGSLEY, S. A., CULSHAW, B., DAVIES, D. E. N., and HALL, T. J.: 'Fibre optic microphones and hydrophones, a comparison with conventional devices' *in* 'Proceedings 1978 international optical computing conference', London, England, 1978 (IEEE, 1978)

7.35 TIMOSHENKO, S. P., and GOUDIER, J. N.: 'Theory of elasticity' (McGraw-Hill, 1970)

7.36 BUDIANSKY, B., DRUCKER' D. C., KINO, G. S., and RICE, J. R.: 'The pressure sensitivity of clad optical fibre', *Appl. Opt.*, 1979, 18, p. 4085

7.37 HALL, T. J.: Ph.D. thesis, University of London, 1980

7.38 NYE, J. F.: 'Physical properties of crystals' (Oxford University Press, 1976)

7.39 SAGNAC, G.: 'L'ether lumineux demontre par l'effet du vent relatif d'ether dans un interferometre en rotation uniforme', *C. R. Acad. Sci.,* 1913, 95, pp. 708–710

7.40 MICHELSON, M., and GALE: *Astrophysics Journal* 61 p. 140 (1925)

7.41 BERGH, R. A., LeFEVRE, H. C., and SHAW, H. J.: 'All fibre gyroscope with inertial navigation sensitivity', *Opt. Lett.,* September 1982, 7, 9, p. 454

7.42 CULSHAW, B., and GILES, I. P.: 'Fibre optic gyroscope', *J. Phys. E,* 1983. 16, p. 5, Jan 1983

7.43 CULLEN, A. L.: private communication

7.44 BERGH, R. A., LeFEVRE, H. C., and SHAW, H. J.: 'All single mode fibre optic gyroscope with long term stability', *Opt. Lett.,* 1981, 6, p. 502

7.45 HECHT, E., and ZAJAC. A., Optics, p. 502; (Addison Wesley 1973) or BALDWIN, G. C.: 'An introduction to non-linear optics' (Plenum Press, 1969)

7.46 EZEKIAL, S., DAVIS, J. L., and HELLWARTH, R.: 'Intensity dependent non-reciprocal phase shift in fibre gyro' *in* 'Proceedings international conference on fibre optic rotation sensors and related technologies', MIT, November 1981 (Springer Verlag, 1982)

7.47 BERGH, R. A., LeFEVRE, H. C., and SHAW, H. J.: 'Compensation of the optical Kerr effect in fibre optic gyroscope', *Opt. Lett.,* 1982, 7, p. 282

7.48 BERGH' R. A., CULSHAW, B., CUTLER, C. C., LeFEVRE, H. C., and SHAW, H. J.: 'Source statistics and the Kerr effect in fibre optic gyroscopes', *ibid.,* 1982, p. 563

7.49 CULSHAW, B., and GILES, I. P.: 'Frequency modulated heterodyne optical fibre Sagnac interferometer', *IEEE J. Quantum Electron.,* April 1982, 18 (4), p. 690

7.50 AUCH, W., and SCHLEMPER, E.: 'Drift behaviour of a fibre optic station sensor using polarisation preserving fibre,' Proceedings First International Conference on Optical Fibre Sensors London April 1983 – supplementary paper. IEE vol. 221

7.51 DAVIS, J. L., and EZEKIAL, S.: 'Closed loop low noise fibre optic rotation sensor', *Opt. Lett.,* 1981, 6, p. 505

7.52 BARNOWSKI, M. K., ROURKE, M. D., JENSEN, S. M., and MELVILLE' R. T.: 'Optical time domain reflectometry', *Appl. Opt.,* 1977, 16, p. 2375

7.53 LIN, S. C., and GIALLORENZI, T. G.: 'Sensitivity analysis of the Sagnac effect optical fibre ring interferometer', *ibid.,* 1979, 16 (6), p. 915

7.54 COBB, K. W., and CULSHAW, B.: 'Reduction of optical phase noise in semiconductor lasers', *Electron. Lett.,* 1982, 18 (8), p. 338

7.55 KOBAYASHI, S., YAMAMOTO, Y., ITO, M., and KIMURA, T.: 'Direct frequency modulation in AlGaAs semiconductor lasers', *IEEE J. Quantum Electron.,* April 1982, 18, p. 582

7.56 See, for instance:
'Proceedings SPIE conference on optical rotation sensing' (SPIE, Vol. 157); or ARONOWITZ, F.: 'The laser gyro', *Laser Appl.,* 1971, 1, pp. 133–200

7.57 See 'Proceedings conference on fibre optic rotation sensing and related technologies', MIT, November 1981 (Springer Verlag, 1982)

7.58 LESLIE, D. H., TRUSTY, D. L., DANDRIDGE, A., and GIALLORENZI, T. G.: 'A fibre optic spectrophone', *Electron. Lett.,* 1981, 17, p. 581

Chapter 8

8.1 DURST, F., MELLING, A., and WHITELAW, J. H.: 'Principles and practices of laser Doppler anemometry' (Academic Press, 1976)

8.2 BACHMAN, C.: 'Laser radar systems and techniques' (Artech, Dedham, Mass., 1979)

8.3 DYOTT, R. B.: 'The fibre optic Doppler anemometer', *IEE J. Microwave, Opt. & Acoust.,* January 1978, 2 (1), p. 13

8.4 TANAKA, T., and BENEDEK, G. B.: 'Measurement of the velocity of blood flow (invivo) using a fibre optic catheter and optical spectroscopy', *Appl. Opt.,* 1975, **14**, p. 189

8.5   TEICH, M. C.: 'Infrared heterodyne detection,' *Proc. IEEE,* **56**, p. 37, 1968

8.6   YARIV, A.: 'Introduction to optical electronics' (Holt, Rinehart and Winston, 1975)

8.7   PIKE, E. R.: 'Photon correlation spectroscopy' *in* CUMMINS, H. Z., and PIKE, E. R. (Eds), NATO Advanced (Plenum, 1977)

8.8   HECHT, E., and ZAJAC, A.: 'Optics' (Addison Wesley, 1973) p. 79

8.9   HUDSON, M. A.: 'Calculation of the maximum optical coupling efficiency into multimode optical waveguides', *Appl. Opt.,* 1974, **13**, p. 1029

8.10 KIAZAND, B.: M.Sc. thesis, University of London, 1978

8.11 ERDMANN, J., and SOREIDE, D. C.: 'Fibre optic laser transit velocimeter', *Appl. Opt.,* June 1982, **21** (11), p. 1876

8.12 NILSSON, G. E., TENLAND, T., and AKEOBERG, P.: 'Evaluation of a laser Doppler flowmeter for measurement of tissue blood flow', *IEEE Trans. Biomed. Eng.,* October 1980, **27** (10), pp. 597–604

8.13 NISHURA, H., KOYAMA, J., KOKI, N., KAJIYA, F., HIRONAGO, M., and KANO, M.: 'Optical fibre laser Doppler velocimeter for high resolution measurement of pulsatile blood flow', *ibid.,* May 1982, **21** (10), p. 1785

8.14 See, for instance: STOLEN, R. H.: 'Fiber design for non-linear optics' *in* 'Physics of fibre optics – advances in ceramics Vol. II' (American Ceramic Society, 1981)

Chapter 9

9.1   DAKIN, J. P.: 'A novel fibre optic temperature probe', *Opt. & Quantum Electron.,* 1977, **9**, p. 540

9.2   FRANCIS, J. J.: 'The design of optical spectrometers' (Chapman and Hall, 1969)

9.3   PETERSON, J. I., GOLDSTEIN, S. R., and FITZGERALD, R. V.: 'Fibre optic pH probe for physiological use', *Anal. Chem.,* 1980, **52**, p. 864

9.4   The fluoroptic thermometer: literature available from Luxtron, 1060 Terra Bella Avenue, Mountain View, CA 94043, USA

9.5   See: *Optical Engineering,* November/December 1981, **20** (6), a special issue on tunable optical spectral filters

Chapter 10

10.1 HECHT, E., and ZAJAC, A.: 'Optics' (Addison Wesley, 1973) p. 257

10.2 BORN, M., and Wolf, E.: 'Principles of optics' (Pergamon Press, 1975) Chap. 14

10.3 ROGERS, A. J.: 'The electrogyration effect in crystalline quartz', *Proc. Roy. Soc. A,* 1977, **353**, pp. 117–192

10.4 YARIV, A.: 'Introduction to optical electronics' (Holt, Rinehart and Winston, 1976)

10.5 NYE, J. F.: 'Physical properties of crystals' (Oxford University Press, 1976)

10.6 ROGERS, A. J.: 'Optical measurement of current and voltage on power systems' *IEEE J. Electr. Powe Appl.,* 1979, **2** (4), p. 120

10.7 NORMAN, S. R., PAYNE, D. N., ADAMS, M. J., and SMITH, A. M.: 'Fabrication of single mode fibres exhibiting extremely low polarisation birefringence', *Electron. Lett.,* 1979, **15**, p. 309

10.8 SMITH, A. M.: 'Optical fibre current measurement device at a generating station' *in* 'Proceedings European conference on optical systems and applications', Utrecht, *in* 1980, 'Proceedings SPIE', Vol. 236, p. 352

10.9 ROGERS, A. J.: 'Polarisation optical effects and their use in measurement sensors' *in* 'Proceedings optical sensor and optical techniques in instrumentation' (Institute of Measurement and Control, November 1981, London, England)

10.10 ROGERS, A. J.: 'An optical temperature sensor for high voltage applications', *Applied Optics* 21, p. 882, 1982

Chapter 12

12.1 GOODMAN, J. W.: 'Introduction to Fourier optics' (McGraw-Hill, 1968)

12.2 GASKILL, J. D.: 'Fourier transforms, linear systems and optics' (Wiley, 1978)

12.3 PRESTON, K.: 'Coherent optical computers' (McGraw-Hill, 1972)

12.4 CASASENT, D.: 'Optical data processing', (Springer Verlag 1978)

12.5 GOODMAN, J. W.: 'Architectural development of optical data processing systems' *in* 'Proceedings CLEO '82', Paper THF1, Phoenix, Arizona (Optical Society of America)

12.6 SMITH, K. C.: 'The prospects of multivalued logic – a technology and applications view', *IEEE Trans. Comput.,* September 1981, 30 (9), pp. 619–634

12.7 See for instance: *Computer*, January 1982, 15 (1), special issue on highly parallel computing

12.8 TAMIR, T.: 'Integrated optics' (Topics in applied physics, Vol. 7, Springer Verlag, 1979, 2nd edn.)

12.9 YARIV, A.: 'Introduction to optical electronics' (Holt, Rinehart and Winston, 1976) Chap. 13

12.10 CUTRONA, L. J., LEITH, E. N., PORCELLO, L. J., and VIVIAN, W. E.: 'On the application of coherent optical processing techniques to synthetic aperture radar', *IEEE Proc.,* August 1966, 54, p. 1026

12.11 ROTZ, F. B.: 'Compact time integrating optical correlator with variable time scaling', *in* 'Proceedings SPIE., 1980, Vol. 241, 'Real time signal processing III', p. 161

12.12 HAZAN, J. P.: 'Application of Tilus and Phototilus to Data Processing.'. Proc. IOCC 1978 (London), p. 17. (IEEE New York)

12.13 GRINBERG, J. W., BLEHA, P., JACOBSON, A. D., LACKER, A. M., MYER, G. D., MILLER, L. J., MARGERUM, J. D., FRAAS, L. M., and BOSWELL, D. D.: 'Photo-activated birefringent liquid crystal light valve for colour symbology display', *IEEE Trans. Electron Devices,* 1975, 22, p. 775

12.14 BARNOWSKI, M. K. *et al.*: 'Design, fabrication and integration of components for an integrated optics spectrum analyser' *in* 'Proceedings 1978 ultrasonics symposium' (IEEE)

12.15 MERGERIAN, D., *et al.*: 'An integrated optic r.f. spectrum analyser', *in* 'Proceedings SPIE', Vol. 239, July 1980, p. 121

12.16 DOUGHTY, G. F., *et al.*: *in* 'Proceedings SPIE', Vol. 235, pp. 35–41

12.17 TRIMBLE, J., CASASENT, D., PSALTIS, S. D., CAIMI, F., CARLOTTO, M., and NEFT, D.: 'Digital correlation by optical convolution/correlation', *in* 'Proceedings SPIE', 1980, Vol. 241, p. 155

12.18 SKOLNIK, M.: 'Introduction to radar systems' (McGraw-Hill, 1962)

12.19 BENJAMIN, R.: 'Modulation, resolution and signal processing in radar, sonar and related systems' (Pergamon Press, 1966)

12.20 JACK, M. A., and GRANT, P. M.: 'Design and applications of surface acoustic wave Fourier transform processors' *in* KUNT, M., and DE COULON, F. (Eds), 'Signal processing, theory and application' (North Holland, 1980)

12.21 PRESTON, K.: 'Coherent optical computers' (McGraw-Hill, 1972)

12.22 TAYLOR, H. F.: 'Fiber and Integrated Optics for Signal Processing', Proc IOCC, 1978 1978 (London) p. 198 (IEEE New York)

12.23 BERGH, R. A., KOTLER, G., and SHAW, H. J.: 'Single mode fibre optic directional coupler', *Electron. Lett.,* 1980, 16, p. 260

12.24 SHEEM, S. K., GIALLORENZI, T. G.: 'Single mode fibre optical power divider: encapsulated etching technique', *Opt. Lett.,* 1979, **4** (29)

12.25 BOWERS, J. E., NEWTON, S. A., SORIN, W. V., and SHAW, H. J.: 'Filter response of single mode fibre recirculating delay lines', *Electron. Lett.,* 1982, **18** (3) p. 110

12.26 JACKSON, K. P., BOWERS, J. E., NEWTON, S. A., CUTLER, C. C.: 'Microbend optical fibre tapped delay line for gigahertz signal processing', *Appl. Phys. Lett.,* July 1982, **41** (2), p. 139

12.27 BLOOM, D. M., WANG, S. Y., and COLLINS, D. M.: '10 GHz bandwidth GaAs photodiodes', CLEO '82', Paper ThE2 (Optical Society of America, 1982)

12.28 See, for example: 'Proceedings CLEO '82', Session WI

12.29 LEONBERGER, F. J.: 'Applications of guided wave interferometers', *Fibre Opt. Tech.,* March 1982, p. 125

12.30 TAYLOR, H. F.: 'Guided wave electro-optic devices for logic and computation', *Appl. Opt.,* May 1978, **17** (10), p. 1493

12.31 LEONBERGER, F. J., WOODWARD, C. E., and BECKER, R. A.: '4-bit 828 megasample per second electro-optic guided wave analogue to digital converter', *Appl. Phys. Lett.,* 1982, **40** (7), p. 565

12.32 MARCATELLI, E. A. J.: 'An optical picosecond gate', *Appl. Opt.,* May 1980, **19**, p. 1468

12.33 HAUS, H. A., KIRSCH, S. T., MATHYSSEK, K., and LEONBERGER, F. J.: 'Picosecond optical sampling', IEEE J. Quantum Electronics *JQE-16*, p. 870 (1980)

12.34 CUMMINGS, R. C.: 'The serrodyne frequency translator', *Proc. IRE,* February 1957, **45** (2), p. 175

12.35 VOGES, E., OSTWALD, O., SCHEIK, B., and NEYER, A.: 'Optical phase and amplitude measurement by single sideband detection', *IEEE J. Quantum Electron.,* January 1982, **18** (1), p. 124

12.36 KING, R. J.: 'Microwave homodyne systems' (Peter Peregrinus, 1978)

12.37 CULSHAW, B., and WILSON, M. G. F.: 'Integrated optic frequency shifter modulator', *Electron. Lett.,* 1981, **17**, p. 135

12.38 HEISSMANN, F., and ULRICH, R.: 'Integrated optic single sideband modulator and phase shifter', *IEEE J. Quantum Electron.,* April 1982, **18** (4), p. 767

12.39 DAVIES, D. E. N., and KINGSLEY, S. A.: 'An optical fibre data collection highway' *in* 'Proceedings electro-optics/laser international '76', Brighton, England, 1976, pp. 64–72 (Kessler Communications)

12.40 CULSHAW, B., BALL, P. R., POND, J. C., and SADLER, A. A.: 'Optical fibre data collection', *Electron. & Power,* February 1981, **27** (2) p. 148

12.41 TUR, M., GOODMAN, J. W., MOSELI, B., BOWERS, J. E., and SHAW, H. J.: Fibre optic signal processor with applications to matrix-vector multiplication and something filtering, *Optics Letters* 7, 9, p. 463 1983

12.42 See, for instance:
       RABINER, L. R., and GOLD, B.: 'Theory and application of digital signal processing' (Prentice-Hall, 1975); or
       OPPENHEIM, A. V., and SCHAFER, R. W.: 'Digital signal processing' (Prentice-Hall, 1975)

12.43 SMITH, K. C.: 'The prospects for multivalued logic – a technology and application view', *IEEE Trans. Comput.,* September 1981, **30** (9), p. 619

12.44 FATEHI, M. T., WASMUNDT, K. C., YEN, C. Y., and COLLINS, S. A.: 'Sequential logic element arrays for use in optical digital processors: applications and limitations' *in* 'Proceedings SPIE', 1980, Vol. 241, 'Real time signal processing III', p. 139

12.45 BORN, M., and WOLF, E.: 'Principles of optics' (Pergamon Press, 1975) p. 329

12.46 See, for example:
       NORMADIN, R., and STEGEMAN, G. I.: 'A picosecond transient digitiser based on non-linear integrated optics', *Appl. Phys. Lett.,* 1 May 1982, **40** (9), p. 759; and

HALBOUT, J. M., and TANG, C. L.: 'Femtosecond interferometry for non-linear optics', *ibid.,* p. 765

12.47 GARITO, A. F., and SINGER, K. D.: 'Organic crystals and polymers – a new class of non-linear optical materials', *Laser Focus,* February 1982

12.48 SMITH, P. W., and TURNER, E, H.: 'Abistable Fabry Perot resonator', *Applied Physics Letters,* **30,** p. 280, 1977

12.49 TAKAHASHI, H., MASUDA, C., SATOH, S., NAIKI, K., and MIYAJI, K. I.: '2 × 2 optical switch and its applications', *IEEE J. Quantum Electron.,* February 1982, 18 (2), p. 210

12.50 ATHALE, O. R. A., BARR, H. S., LEE, S. H., and BARTHOLOMEW, B. J.: 'Digital optical processing' *in* 'Proceedings SPIE', 1980, Vol. 241, 'Real time signal processing III', p. 149

Chapter 13

13.1 See, for example: 'Proceedings 8th European conference on optical communications', Cannes, September 1982

13.2 DYOTT, R. B.: 'Elliptically cored polarisation holding fibre' in Fibre Optic Rotation Testing and Related TEchnologies,' S. Ezekial and H. J. Ardithy (eds), Springer Verlag 1982

13.3 KAMINOW, I. P.: 'Polarisation in optical fibres', *IEEE J. Quantum Electron.,* January 1981, 17, pp. 15–22
See also: *ibid.,* April 1982, 18, pp. 477–503

13.4 See, for instance: 'Proceedings of the Topical Meeting on Integrated Optics', Asilomar, Ca, January 1982 (Optical Society of America)

13.5 ALFERNESS, R. C., BUHL, L. L., and DIVINO, M. D.: 'Low loss fibre coupled directional coupler modulator', *Electron. Lett.,* June 1982, 18 (12),

13.6 KOBAYASHI, S., and KIMURA, T.: 'Optical FM signal amplification by injection locking and resonant type laser amplifiers', *IEEE Journal of Quantum Electronics,* QE-18, 4, p. 575, April 1982.

13.7 KOBAYASHI, S., and MINURA, T.: 'Optical FM signal amplification by injection locked and resonant type semiconductor and laser amplifiers', *IEEE J. Quantum Electron.,* April 1982, 18 (4), p. 575

13.8 See, for instance: 'K.M.I. market report on fibre optic sensors' (Kessler Marketing Intelligence, 1981); or
'Fibre optics in process control' (Gnostic Concepts Inc., 1981)

13.9 SCHWARTZ, M.: 'Information transmission, modulation and noise' (McGraw-Hill, 1980)

13.10 BOND, D. F.: 'Some practical aspects determining the industrial growth of optical sensors' *in* 'Seminar on optical sensor and optical techniques in instrumentation', London, 12 November 1981 (Institute of Measurement and Control)

13.11 CULSHAW, B.: 'Optical sensor multiplexing systems' *in* 'Proceedings sensors and systems '82', Vol. 1, p. 47 (Network Publications, Buckingham)

13.12 ROGERS, A. J.: 'Optical methods for the measurement of voltage and current at high voltage', *Opt. & Laser Tech.,* December 1977, pp. 273–283

13.13 SKOLNIK, M. I.: 'Introduction to radar systems' (McGraw-Hill, 1963) p. 86

13.14 ROGERS, A. J.: 'POTDR, a technique for the measurement of field distributions', *Appl. Opt.,* 1981, 20 (6), pp. 1060–1074

13.15 ASAWA, C. K., YAO, S. K., STEARNS, R. C., MOTA, N. L., and DOWNS, J. W.: 'High sensitivity fibre optic strain gauge for measuring structural distortion', *Electron. Lett.,* May 1982, 18, p. 362

13.16 KOO, K., and SIGEL, G. H.: 'Fibre optic magnetic sensors based on magnetic glasses', *Optics Letters* 7, p. 334, 1982

13.17 LAGAKOS, N., SCHNAUS, E. U., COLE, J. H., JARZYNSKI, J., and BUCARO, J. A.: 'Optimising fibre coatings for interferometric acoustic sensors', *IEEE J. Quantum Electron.,* April 1982, 18 (4), p. 683

13.18 AL CHALABI, S.: unpublished work

13.19 See, for instance: MAINS, J. D., and PAIGE, E. G. S.: 'Surface wave acoustic devices for signal processing applications', *Proc. IEEE,* May 1976, 64 (5), p. 639

13.20 US Patent 4071753

13.21 CULSHAW, B., GILES, I. P., POND, J. C., and SADLER, A. A.: 'The "Fibredyne" data collection system for industrial telemetry applications' *in* 'Proceedings SPIE', Vol. 355, paper 20

Appendix

A.1 BORN, M., and WOLF, E.: 'Principles of optics' (Pergamon Press, 1975)

A.2 HECHT, E., and ZAJAC, A.: 'Optics' (Addison Wesley, 1973)

A.3 Course on images and information, ST291 Units 1–10 (Open University Press)

A.4 YARIV, A.: 'Introduction to optical electronics' (Holt, Rinehart and Winston, 1976)

A.5 GOODMAN, J. W.: 'Introduction to Fourier optics' (McGraw-Hill, 1968)

A.6 MANDEL, L., and WOLF, E.: 'Coherence properties of optical fields', *Rev. Mod. Phys.,* April 1965, 37 (2) p. 231

A.7 FRANCON, M.: 'Optical interferometry' (Academic Press, 1966)

A.8 Bracewell, R.: 'The Fourier transform and its applications' (McGraw-Hill, 1980)

A.9 YARIV, A.: 'Quantum electronics' (Wiley, 1975)

A.10 KOGELNIK, H., and Li, T.: 'Laser beams and resonators', *Proc. IEEE,* 1966, 54, p. 1312

A.11 KOGELNIK, H.: 'On the propagation of Gaussian beams of light through lens-like media including those with loss and gain variations', *Appl. Opt.,* 1965, 4, p. 1562

A.12 NYE, J. F.: 'Physical properties of crystals' (Oxford University Press, 1976)

A.13 JONES, R. C.: 'A new calculus for the treatment of optical systems', *J. Opt. Soc. Am.,* 1941–48, 27, 31, 32, 38

A.14 AZZAM, R. M. A., and BASHARA, N. M.: 'Ellipsometry and polarized light', North Holland, 1977

A.15 BORN, M., and WOLF, E.: 'Principles of optics' (Pergamon Press, 1975) p. 31

Index